The Astronomers' Library

The Astronomers' Library

The Books that Unlocked the
Mysteries of the Universe

KAREN MASTERS

IVY PRESS

Quarto

First published in 2023 by Ivy Press
an imprint of The Quarto Group.
One Triptych Place, London, SE1 9SH
United Kingdom
T (0)20 7700 6700
www.Quarto.com

A catalogue record for this book is available from the
British Library.

ISBN 978-0-7112-8981-9
EBOOK ISBN 978-0-7112-8983-3

10 9 8 7 6 5 4 3 2 1

Cover Designer: Daisy Woods
Design and Editorial: Roland Hall
Editorial Director: Jenny Barr
Publisher: Richard Green
Senior Production Controller: Rohana Yusof

Printed in China

CONTENTS

INTRODUCTION

Astronomy is both one of the most ancient sciences and a supremely modern science. Since the earliest times, humans all over the world have gazed at the skies, and some noticed and recorded patterns in the motion or appearance of the stars which they connected to their lives on Earth. Marking the passage of the seasons, noticing how the tides followed the phase of the Moon and the ever-present daily motion of the Sun, the impact of the heavens down on Earth was clear. The dates of many of our religious festivals and even the names of the seven days of our week are influenced by this early preoccupation with astronomy. Today highly sophisticated telescopes on the top of tall mountains – even launched into space – continue this tradition of humanity's search to understand the universe we live in.

It is a beautiful and fascinating universe, filled with galaxies reaching to unimaginable and ever-increasing distances, scattered about in a web-like pattern which reveals details of how the universe itself formed. Each galaxy contains millions of stars, each a giant sphere of hot hydrogen and helium with a nuclear furnace at its core, and there is now evidence that every star likely carries with it a family of planets. We've discovered an 'invisible' universe of interstellar gas revealed by radio waves or X-rays, and even dark matter revealed only by its gravitational impact. We have found extreme objects resulting from the death of stars: white dwarfs, neutron stars and black holes, which boggle the mind with their intense gravity and exotic physics. We have even found supermassive

RIGHT

Library study

The author is shown at work, teaching a class in 2019 in the Haverford College's Strawbridge Observatory Astronomy Library.

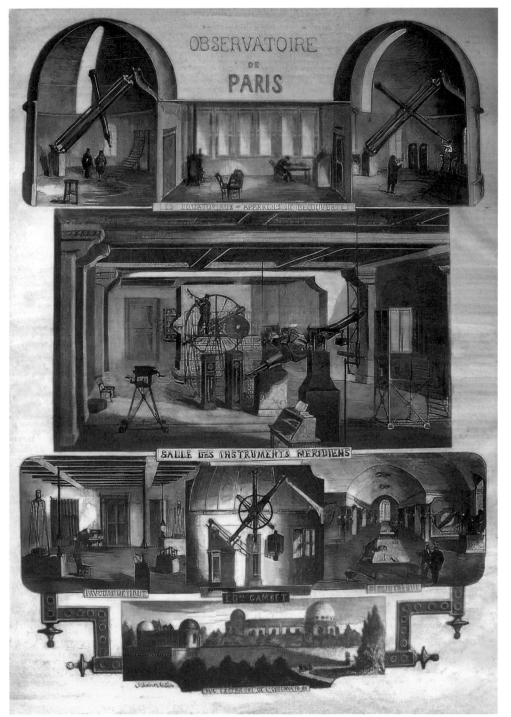

LEFT

Observatoire de Paris

A historic plan from
the Paris Observatory
showing 'La Bibliothèque',
which they claim is one
of the most beautiful
astronomical libraries in
the world.

Learning evolves

The front cover of Galileo's *Dialogo* from 1632, described more in Chapter 4, Developing our Model of the Universe.

The MUL.APIN clay tablet dating from 1,000–500 BCE. This is an early star catalogue - listing 66 different stars and various constellations and also including information about rising times and dates of various stars, as well as describing the path of the Moon.

A seven-prism spectroscope, made by John Browning in 1868 for the astronomer Sir Norman Lockyer. Lockyer used it to split the light from the Sun into a detailed rainbow, and discovered a previously unknown spectral line which revealed a new element in the Sun – he called it 'Helium'.

The Klementinum Library in Prague, Czech Republic. It is home to many classics of Czech literature, as well as astronomy texts.

black holes in the centre of every galaxy in the universe and have detected ripples in the fabric of space itself, caused by the gravitational eddy from merging black holes. Closer to home, the planets in our own Solar System are diverse and make exciting potential destinations for space travel and more: from mighty Jupiter with a host of fascinating moons and a storm system larger than Earth itself, to tiny Mercury, orbiting so close to the Sun that spotting it at dawn or dusk remains a challenge. Smaller members of the Solar System, dwarf planets such as frosty Pluto, or asteroids, comets and shooting stars (meteorites) also hold their charm. Images and drawings of all these celestial bodies capture our imagination. No wonder books about astronomy are so popular.

I have been fascinated by astronomy since a fairly young age, and libraries have also always been special places for me, offering another opportunity to discover worlds beyond the mundane and ordinary. It is common in the 21st century to lament the imminent death of libraries, with the increasing digitization of materials and internet searches bringing vast quantities of human knowledge to our fingertips. However, libraries have a long history in civilization and have constantly evolved to suit the world around them, so I have no doubt they will remain indispensable for many years to come. The importance of free and open libraries to bring books and other materials, astronomical or otherwise, to people who might otherwise have no access to them can never be overstated.

The first libraries contained collections of the most important human knowledge of the time, likely only

accessible to the most learned few. There is evidence of libraries full of clay tablets in ancient Mesopotamia, manuscripts of papyrus in Ancient Egypt, and silk scrolls in Han dynasty China. During the Islamic golden age, following the invention of paper, libraries flourished, filled with numerous priceless copies of earlier and contemporary manuscripts; much of our knowledge of the classical world was saved to the modern era by the Islamic practice of manuscript copying. In medieval Europe, libraries first appeared in monasteries and later in the first universities; these early European libraries chained books to the shelves because they were so precious, although sometimes books could be borrowed for a fee. It was the innovation of movable type and the printing press which really made books available to the masses, and the first recognizable public lending libraries started to become common by the 18th century. In the USA, Benjamin Franklin founded the first free public library in Philadelphia, not far from where I am writing. Today almost every town has a public library, they are found in schools, universities and observatories.

An astronomers' library is, at its simplest, any collection of books about astronomy. Astronomy books don't represent a large fraction of all books ever published so they tend to be a small fraction of any general collection. In a publication of the Pomona College Astronomical society from 1916, a researcher named Elma Schowalter reported on data she collected from a survey of public libraries in California which found that only just over one tenth of a percent of all books in those libraries were about astronomy. Several of those libraries contained no books on the subject at all. So an astronomers' library remains a rare and special place.

Tycho Brahe's Uraniborg, the last major observatory built before the invention of the telescope (circa 1580), may have contained the first genuine astronomical library. Brahe also included a printing press and a paper mill as two of the important astronomical instruments to be contained in the observatory, a clear signal of the importance he placed on being able

to create books about astronomy. Many of the astronomical observatories which were built in the years following, as well as being designed as places to hold telescopes, also hold collections of astronomical books, or astronomical libraries. The observatory in which my office is located, Strawbridge Observatory at Haverford College in Pennsylvania contains an octagonal astronomical library with over 3,000 items, mostly books about astronomy, collected to teach the next generation of college students and support astronomy faculty research. Much larger astronomy libraries are found at larger observatories, designed to serve the researchers who use them, and preserving a precious collection of historical knowledge. To name just a few, there are large astronomical libraries at the Greenwich Observatory in London, the Cambridge Observatory, L'Observatoire de Paris, and the Wolbach Library at Harvard College Observatory. In Italy, the libraries of the National Institute for Astrophysics (INAF) contain over 125,000 items. And, while it is not in an observatory, I can't fail to mention the fabulous Royal Astronomical Society library in London with over 25,000 bound volumes, including some which date back to the 15th century. In this book we curate just a tiny fraction of all the books ever printed about astronomy in our own small 'Astronomers' Library'. This library is organized into topical sections, within which books appear roughly chronologically. We start with a section on Star Atlases (Chapter 1). While these are not always published in books, many beautiful examples exist throughout history and they form an important and eye-catching record for astronomers.

Next comes a selection of books containing maps of other worlds (Chapter 2), ordered initially by object, starting with the Sun and Moon (including eclipses), maps of other planets, (it seems Mars has inspired the most books), and a section on comets. Our collection ranges from the first maps to be drawn using telescopes to modern observations with robotic space explorers.

Astronomy as a science has always had an outsized impact on human culture. In Chapter 3, Astronomy and Culture, we collect historical examples of astronomical knowledge from all over Earth, including the Islamic golden age, the Keralan School of Mathematics, the astronomical knowledge of pre-Columbian Mesoamerica, medieval Europe and East Asia.

Chapter 4 traces humanity's efforts in 'Developing our Model of the Universe', from the geocentric Solar System to the heliocentric revolution and beyond. It includes important steps in the discovery of physical models for gravity and how it impacts the motions of the planets as well as the discovery of the rest of the universe, including galaxies external to our own and the expanding universe.

Most of the books ever published on astronomy were published with a goal of education. In Chapter 5, Astronomy for Everyone, we collect examples of educational books on astronomy, many of which are charmingly illustrated. These range from ancient texts published in medieval Europe through to books containing early astronomical photography at the beginning of the 20th century.

We conclude with a selection of modern books on astronomy – published in the last 100 years or so. This list is not intended to be complete, but unapologetically contains a selection of notable books, either as bestsellers or author favourites.

Karen Masters, Haverford, June 2023

STAR ATLASES

S tar maps form some of the most beautiful printed records of astronomy through the ages. They amaze us with their splendour, variety and beauty. So, it is fitting to start our Astronomers' Library with a tour through star maps across time and from different cultures across Earth.

The first star maps did not come in what we would recognize as books today. Records of our ancestors' fascination with the night skies – even in prehistory – are found in cave drawings, carvings and other objects. There are quite a few claims to the earliest ever star map. One highly disputed claim suggests that a 32,000-year-old Ach Valley tusk fragment shows a representation of the constellation Orion. Several other prehistoric cave art drawings of animals or people have been suggested as possible early constellation art. From around the 7th century BCE, there is the MUL.APIN (or the Plough/Big Dipper), a clay tablet record of Babylonian astronomy and astrology which includes a star catalogue listing constellations.

Another famous (and also debated) example of an early star atlas, or at least a representation of the night skies, is the Nebra Sky Disc (pictured overleaf), made by Early Bronze Age Europeans around 1600 BCE. It shows images interpreted to be the Sun or full Moon, a crescent Moon and the Pleiades (a collection of seven stars). Other early star maps of note from the ancient world are mostly carved or engraved in stone.

Ancient Greek mythology says that Atlas (the Titan after whom, of course, all collections of maps are named) was tasked with holding up the heavens for all eternity.

RIGHT
Dunhuang Star Chart

On of the earliest known graphical representations of the stars in ancient Chinese astronomy. The Dunhuang Star Chart is dated to Tang dynasty China (618–907 CE) and was part of a collection of manuscripts discovered in Dunhuang, China in 1900.

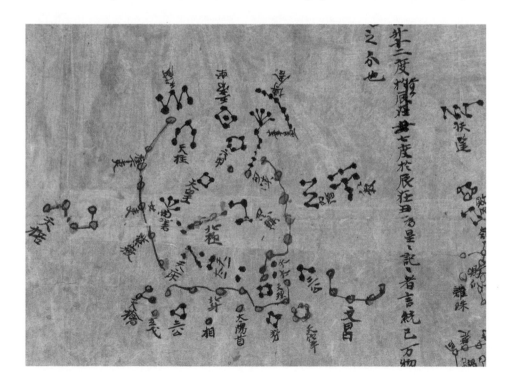

Farnese Atlas

Dating from the 2nd century, this marble sculpture depicts the Titan Atlas holding up a celestial globe, which is engraved with one of the earliest known depictions of the traditional ancient Greek constellations. The statue is housed in the National Archaeological Museum of Naples in Italy.

Cheongsang Yeolcha Bunyajido

Dating from the 14th century, this star map shows 1,467 stars that were visible from Korea at the time. It entered standard usage in Korea and was used until the 19th century, when Western planispheres became more popular.

Statues of Atlas often feature a celestial sphere, decorated with representations of the stars and constellations. The oldest known, the Farnese Atlas, dated to 150 BCE, is also the earliest known representation of the classical (Greek) constellations, which likely date from many hundreds, if not thousands of years earlier.

Constellations are dot-to-dot pictures humans make with the stars, so across human history, geographically separate cultures naturally developed different constellation patterns, even though they all observed the same stars. Another beautiful early engraved stone star map is the Cheonsang Yeolcha Bunyajido, from 14th-century (Josean dynasty) Korea, showing the constellations from Korean culture.

In the medieval Arab world, astrolabes were particularly popular, and since they were often made from metal, many exist to this day containing records of the constellations before printed maps. An astrolabe has multiple metal discs which can rotate against each other, allowing the experienced user to perform calculations of an Earth-bound position from the locations of the stars shown on the object.

One of the earliest known paper star charts can be found in a Chinese scroll known as the Dunhuang Star Chart. It dates back to the Tang dynasty (7th and 8th centuries). The Chinese tradition divided stars into 257 constellations in 12 segments (one per month of the year). The scroll survived hundreds of years in a desert cave, until its discovery in the early 20th century, bringing us a priceless record of early Chinese constellations.

ABOVE LEFT AND RIGHT
Sky Disc and astrolabe

The *Himmelsscheibe von Nebra* (or Nebra Sky Disc) was found near Nebra in central Germany and dates from the Bronze Age. The symbols are thought to represent stars (possibly the Pleiades cluster), a lunar crescent and a full Moon or Sun.

Planispheric astrolabe: Allowing a user to calculate positions of the stars, astrolabes were often made of metal and therefore many have been preserved. This example, dating from the 9th century, is cast from brass and was found in North Africa.

The Book of the Images of the Fixed Stars

One of the oldest surviving illustrated books, *The Book of the Images of the Fixed Stars*, or sometimes *The Book of Constellations* is a complete atlas of the night sky as known in the Arabic world during the 10th century. The author, Abd Al-Rahman Al Sufi (903–986), was born in Persia (in modern-day Iran) in 903. Al Sufi (sometimes called Azophi in historical texts from the West) was an astronomer famous for his work mapping the night skies.

While the original copy of the book no longer exists, thanks to the Islamic tradition of transcribing manuscripts, there is more than one almost contemporaneous copy. Al Sufi must have been familiar with both the Ancient Greek traditions for the constellations as well as the Arabic (likely Bedouin) traditions, and presented them in this book, a beautiful merger of the two. In addition to the illustrations of the constellations, the book contains precise astronomical coordinates for the stars. Because Earth's axis shifts very slightly over time, making a complete circle (or precession) in 26,000 years, the rise and set times of stars shift gradually, and calculations of their coordinates are only precise for a short period of time. *The Book of the Images of the Fixed Stars* gives their positions calculated to be exact in the year of 964.

Of note, each Greek constellation is presented in this book with two separate illustrations – one presented as it would appear in the sky, and the second in a mirror image view suitable for use by makers of celestial globes (showing an 'outside in' view of the skies). Another notable feature of the book is the first ever record of a well-known (to Arabic astronomers) 'little cloud', in the Andromeda constellation (or the Arabic Big Fish). This object is now understood to be the nearest large galaxy to our own, and the only galaxy visible to the unaided eye in the northern hemisphere.

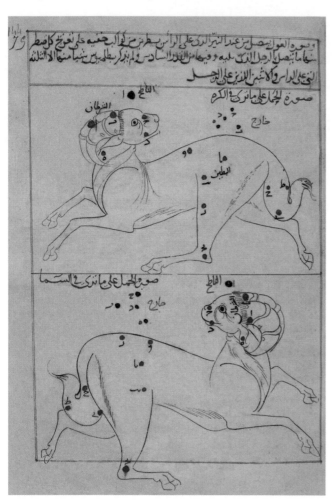

The Greek constellation, Andromeda (the daughter of Cassiopeia who is saved from a sea monster by Perseus) is also the Arabic constellation the Big Fish. The 'little cloud' that we know today as the Andromeda galaxy is marked by the collection of black dots on the mouth of the fish.

The Book of the Images of the Fixed Stars

The top two images show
the two mirror image
views of the Ursa Major
constellation (from
different editions of the
book). The one facing to
the right is how it appears
in the sky, the other – for
use in making a celestial
globe – is from the oldest
existing edition of the
book dating to 1009.
The book also included
tables of star positions
calculated for the year
of 964.

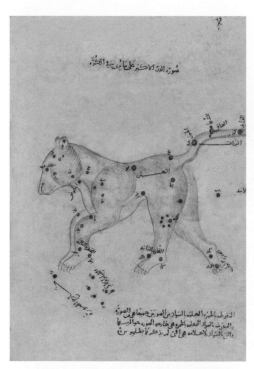

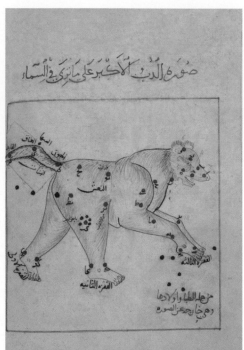

Further examples of the Andromeda and/ or Big Fish of Arabic constellations), although the marks showing the location of Andromeda seem to be missing from this version. In the mirror image version (opposite) even the fish is missing.

Poeticon astronomicon

Poeticon astronomicon, or Poetic Astronomy, is a book of retellings of the Ancient Greek and Roman constellation myths, accompanied by illustrations of the star patterns and constellation art. It was written by Casius Julius Hyginus and was first published 1482 in Latin. The first publication came well over 1,000 years after it was supposed to have been written, and there exists some debate as to the true authorship of this medieval work. Some scholars claim the level of language in the Latin is too simple for Hyginus (64 BCE–17 CE) to have been the true author, since Hyginus normally wrote in a more complex style. The constellations appear in the same order as in Ptolemy's *The Almagest*, which was published after Hyginus died.

Whatever its origin, it provides a record of Greek mythology around the constellations, alongside charming medieval illustrations of the figures.

RIGHT AND OPPOSITE
Poetic Astronomy

The constellations of Draco, Ursa Major and Ursa Minor are shown in this hand-coloured woodcut from a later version of the book.

A number of constellations (Cancer, Leo and Cassiopeia) are easily identified in the black and white edition of the book (with hand-coloured, illuminated lettering).

Ancer. Hunc medium diuidit circulus æstiuus ad leonis exortus spectantē: paululũ supra caput hy/ drę collocatum. Occidentē & exorientem poste/ riore corpis parte. Hic autē habet in ipsa testa stel/ las duas quę asini uocant: de quibus ante diximꝰ. In pedibꝰ dextris singulas obscuras. In sinistro pede ꝑmo duas In secūdo duas obscuras. In tertio unam. In quarto primo umā obscuram. In ore unam. In ea quę chela dexterior dicīt tres si/ miles: nō grandes. In sinistra similes duas. Et ita ē omnino stel/ larum numerus decem & octo.

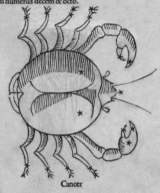

Cancer

 Eo spectās ad occasum supra corpus hydrę a capi/ te qua cancer īnstatuīsꝗ ad mediā partē eius con/ stituīt medius æstiuo circulo diuiditur: ut sub ipso orbe priores pedes habeat collocatos. Occidens a capite & exoriēs. Hic habet in capite stellas tres In ceruicibꝰ duas. In pectore unam. Interscapilio tres. In media cauda unā. In extrema alterā magnā. Sub pectore duas. In pe/ de priore unā clarā. In uentre clarā unam. Et infra alterā ma/ gnam. In lumbis unam. In posteriore genu unam. In pe/ de posteriore claram unam. Et ita est omnino numerus stellarũ decem & nouem.

Leo

 B hoc circulo abest circulus tonon dimidiũ: quo loco Mercurii sidus uehitur. Itaꝗ diebꝰ. xxx. ad alterum signum transiens. Tardius ab hoc circulo abest tonon dimidiũ: quo loco iter suum ueneris dirigit astrum. Tardius ꝗ Mercurii stella consi/ ciens curtũ. Trāsit enim ad aliud signũ diebus. xxx. Supra hu/ ius stellam solis est cursus qui abest ab hespero: quę est ueneris stella medietate toni. Itaꝗ cum inferioribus pariter peruolans uno anno idest duodecim. signa pcurrit tricesimo die ad aliud transiens signum. Supra solem igit & eius circulum Martis est stella quę abest a sole tono dimidio. Itaꝗ dicitur diebus. lx. ad aliud signũ transire. Hunc orbem supra Iouis ē stella: quę abest a martis hemitonio. Itaꝗ uno anno ad alterum transit signum. Nouissima stella saturni quę maximo uehit circulo: hęc autem tono distat a Ioue. Itaꝗ annis. xxx. xii. pcurrit signa. Ab Ipsorũ tamen siderum corporibus Saturnus abest tono uno & dimi/ dio. Hac igitur ratione potes scire neꝗ solem neꝗ lunam con/ tingere stellas: & nihilominus p zodiacũ circulum uerti. Hinc etiam possumus intelligere: lunam minorē eē sole. Omnia quę proxima sunt nobis maiore necesse est esse: ꝗ quę longo disce/ dente interuallo uidemus. Igitur lunā uidemus proxime nos esse. Neꝗ eam maiorē nostro aspectui esseꝗ solem. Illud quoꝗ necesse est cum sol nō longe ablit a luna: & a nobis maior uide/ tur. si prope accesserit: multo maiorē futurum. Pręterea necesse est ut ante diximus: aut nullam stellam erraticā esse: aut solem pariter cum luna & reliquas stellas errare. Si enim quisꝗ mihi potest demonstrare quinꝗ stellarum cursum & diceret: ꝗ bodie quęꝗ coram ad aliud transeat signũ: quēadmodũ de sole & lu/ na fieri uidemus: & nihilominus suũ efficit cursum nō est erra/ tica. Si autē dubium est ꝗ bodie transeat: & ad aliud signũ cō/ pari ratione cum luna feratur. & suũ circulum dirigat: quemad/ modum stellę quę stĩt dubię. necesse est has quoꝗ errare. sed nō

 Assiopeia sedes in siliquastro collocata ē. cuius sedi/ lis & ipsꝯ cassiopeię pedes positi in ipsa circulu/ ctione circuli qui arcticus uocat. effugiens autē cor/ poris ad æstiuũ circulũ puenit: quę capite & dextre ra manu tāgit. Hác ꝓpe mediā diuidit circulus is lacteus appellat. ꝓxime cephei signũ collocatũ. Hęc occidēs cũ scorpione capite: cũ sedili resupino ferri pspicit. Exoriri autē cũ sagictario. Huꝯ in capite stella ostendif una. In utroꝗ humero una. In mamilla dextra clara una. In lumbis magna una. In sini/ stro femore duas. In genu una. In pede ipso dextro una. In qua/ drato quo sella deformat una. In utrisꝗ singule clarī: ęteris lu/ centes. Hęc igīt est oīno stellarũ. xiii.

Cassiopeia

De le stelle fisse

Widely regarded to be the first printed star atlas (book of maps of constellations), *De le stelle fisse* (The Fixed Stars) by the Italian Renaissance scholar Alessandro Piccolomini (1508–1579) is a book of maps of almost all of the traditional Greek constellations.

Unlike earlier all-sky maps, or many later star atlases, the drawings are quite plain. They are not necessarily oriented North-South, but rather displayed in orientations which make it easier for the astronomer to use them to identify stars in a constellation. Piccolomini also wrote in everyday Italian (rather than Latin) likely with a goal of providing an accessible astronomical text, and while no artwork was included, each constellation map was accompanied with a description of myths associated with it. Of note to astronomers today, Piccolomini was the first to use a lettering system for stars in a constellation with 'a' denoting the brightest star and so on. He also included tables of the 'magnitude' (a ranking of brightness) of stars from 1st (brightest in the sky) to 4th.

Piccolomini was born into a prominent and scholarly family in Siena, an important Italian city state in medieval and Renaissance Europe. As well as his two books on astronomy *De le stelle fisse* and *De la sfera del monde* (On the Sphere of the World), Piccolomini published poetry, philosophy and plays, and was well known for translating works from the Classics into everyday Italian. He appears to have had a long-standing infatuation with the Italian poet Laudomia Forteguerri, who is supposed to have inspired his astronomical works (which are dedicated to her) after lamenting that as a woman she was not able to study astronomy. In this way his book may have been the first astronomy book aimed at opening up astronomy to the general public (including women).

LEFT

De le stelle fisse

Clockwise from top left:
title page of the 1579
edition; an engraving
of the author from
another publication; the
constellations of
Ursa Minor and Ursa
Major from the book.

Uranometria

The *Uranometria* was named after either Urania, the Greek muse of astronomy, or Ouranus, the Greek personification of the heavens and husband of Gaia, or maybe both; the planet Uranus was not discovered until almost 200 years after the 1603 publication of this book. It represents the first of the great pictorial sky atlases in our Astronomers' Library, and the first to include maps of the entire night skies. The author, Johann Bayer (1572–1625), trained and worked professionally as a lawyer in Germany, but he is most famous for his interest in astronomy and his publication of *Uranometria*.

Bayer's work built on the Piccolomini atlas (*De le stelle fisse*) published 60 years earlier (see previous pages), but included additional fainter stars in the familiar Greek constellations, as well as completely new areas of the skies – parts which are not visible from the northern hemisphere so were unknown to the Ancient Greek astronomers. Bayer also shifted from the Roman lettering for stars used by Piccolomini to the Greek letters that are still in use today to indicate the brightness ranking in a constellation. For example, Betelgeuse, the brightest star in the Orion constellation, is referred to as Alpha Orionus in this system.

Each constellation is illustrated using artwork by Alexander Mair (c.1560–1617) one of the 16th-century pioneers in the use of copper-plate engraving for illustrations. Mair's skill and expertise in this technique enabled the constellation artwork to be both beautiful and precise, and they went down in history as some of the most iconic images

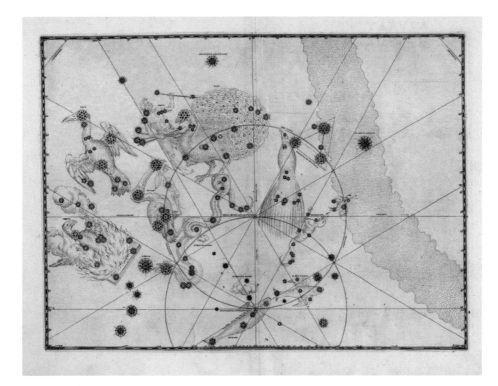

Uranometria

Written by Bayer and
published in Augsburg,
Germany by Christoph Mangle
in 1603. Alexander Mair's
beautifully detailed engravings
built on previous works in
terms of detail as well as
intricacy. Constellations shown
here are Ursa Major (opposite),
Orion (below left) and the 12
constellations of the southern
polar skies, in print in a single
chart for the first time (left).

Uranometria

Lavishly coloured versions of the book have been produced frequently. These images, which date from a 20th-century edition, show the constellations Equuleus, the 'little horse' as illustrated, and Delphinus, the dolphin, shown just as a collection of stars (opposite) and Serpens (below), shown as a complete snake. Modern maps split this into the head and the tail and label the central part 'Ophiuchus, the Serpent Bearer'. Notice also the plane of the Milky Way (our Galaxy) crossing the tail of Serpens.

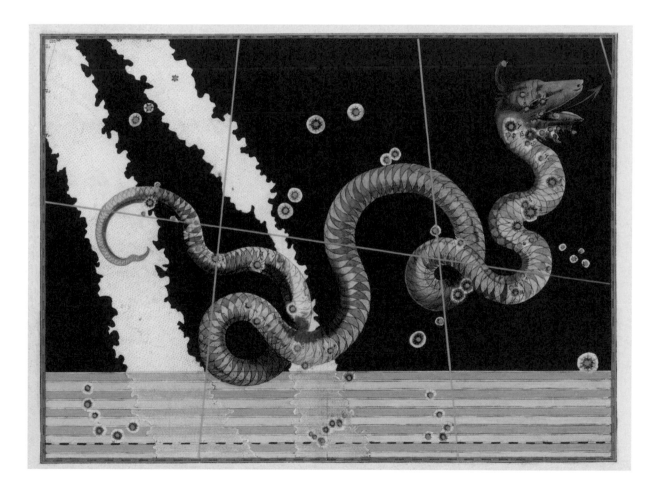

of the constellations. The charts in the *Uranometria* are remarkably accurate. Coordinate grids were overlaid so that positions can be accurately determined and the star positions are in large part based on the measurements given by Tycho Brahe (1546–1601), which Kepler would use to develop his empirical models of the motions of the planets (see Chapter 4: Developing Our Model of the Universe). This is a book which was both beautiful and could be used for state-of-the-art science of the day.

A big innovation in the *Uranometria* was the inclusion of the 12 new constellations in the southern skies. Unless they had travelled significantly, astronomers based in Europe would have never seen the south polar part of the sky, which never rises above the horizon north of the Equator, so this map must have been fascinating to them – almost like exploring a new world. Bayer included all 12 constellations in a single chart of the south polar skies, basing this work on a globe of the sky by the Dutch astronomer Petrus Plancius (1552–1622). Plancius trained a Dutch seafaring navigator (Pieter Keyser, 1540–1596) to chart the skies as he sailed; when the information returned to him, he invented constellations for the patterns of stars. These constellations, named mostly after new animals encountered by Europeans in their southern explorations, are still mostly the 'official' constellations used today, and since Bayer's printed book was more widely accessible than the globe by Plancius, he is often erroneously credited with inventing them.

Uranometria

Many editions of the work have been produced, but the original, black and white editions are still stunning in their detail – and accuracy. This shows the constellation of Aquarius, the water carrier.

Harmonia macrocosmica

The *Harmonia macroscosmica* is considered one of the masterpieces of the so-called 'golden age of Dutch cartography' in which multiple publishers located in the Netherlands competed to produce truly enormous and beautiful multi-volume atlases, which were purchased by rich merchants as status symbols. The author, Andreas Cellarius (c.1596–1665), was a Dutch cartographer and astronomer best known for his work on the *Harmonia macrocosmica*, published in Amsterdam in 1660. Details of his life are not well known. It was likely he was born in Poland, and it has been documented that he worked as a schoolteacher for a while in the Netherlands, as well as having an interest in the skies.

Developments in printing, and the inclusion of colour, along with the inclusion of (to modern eyes) quite whimsical additions in both the main body of and most notably the borders of the illustrations make the plates beautiful. It is likely they are familiar to most astronomers, even if they may have never looked through the book in its entirety.

This book is described by some as the most beautiful celestial atlas ever produced. Published roughly 60 years after the *Uranometria* (see pages 30–34), it was intended to be the final volume in the massive work *Atlas Major*, by Johannes Janssonius (1588–1664). This large-scale, multi-volume atlas was planned by the famous map maker Gerard Mercator (of the Mercator projection) to contain maps of the entire known universe at the time including mostly geographical maps, but also maps of the heavens.

In total, *Harmonia macrocosmica* contains 29 colourful plates of illustrations. As well as the detailed and beautiful all-sky star charts showing various perspectives and illustrations for the constellations, many of these plates contain quite complex, but still beautiful diagrams laying out the world systems (or cosmologies) under debate at the time, including the geocentric model of Ptolemy and the heliocentric, Copernican model (see Chapter 2: Mapping Other Worlds). There is also a charming illustration explaining the reason for the phases of the Moon, and illustrations of the daily and annual motions of the stars and planets in the skies in different models.

BELOW

Harmonia macrocosmia

This page is described (in Latin) as the 'Planisphere of Ptolemy, or the mechanism of the heavenly orbits', showing Ptolemy's traditional geocentric model for the universe.

Harmonia macrocosmia

This image is a depiction of
Ptolemy's geocentric model,
showing Earth surrounded by
the constellations of the zodiac
in the sphere of fixed stars,
within which move the other
planets and the Sun.

Stars of the northern celestial hemisphere

Many editions of *Harmonia macrocosmia* have been produced, some ornately coloured. In this illustration we see a map of the northern skies with traditional Greek constellations.

Harmonia macrocosmia

The edges of the maps are sometimes as interesting as the maps themselves. For example, in the detail in the plate on the right, astronomers are viewing the skies using Jacob's staff or a cross-staff, both from the ground and the tops of nearby buildings. This was an early astronomical instrument used to accurately estimate angles. Nearby, other astronomers help to record the measurements, and a globe is also being used for charting. For some reason the illustration shows them doing this under cloudy daytime skies – which of course would not be possible. Meanwhile, in the clouds above, the cherubs are also having a go with their own cross-staffs!

Multiple beautiful plates like these were devoted to an attempt to move on from the Greek constellations to constellations based on stories from the Bible.

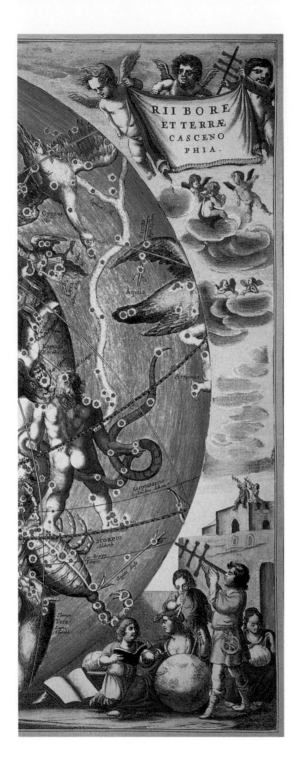

Harmonia macrocosmia: selenographic phases of the Moon.

The Sun at the top shines its rays down, and as the Moon orbits Earth (centre) different parts are illuminated, creating the crescent and other phases seen from Earth.

Firmamentum sobiescianum sive uranographia

A huge technological step occurred for astronomy in the early 1600s with the invention of the telescope. First used by Galileo Galilei to observe the skies in 1609, by the time *Firmamentum sobiescianum sive uranographia* was published, almost 100 years later, telescopes were widely in use in professional astronomy.

One hold-out in this world was Johannes Hevelius (1611–1687), a Polish politician and astronomer. Despite other astronomers having already made the shift, Hevelius believed that using a telescope to map the locations of stars might introduce distortions into his results and so adamantly refused, even including in the edges of his maps comments on how telescopic observations were not as good. He made detailed observations of the positions of stars for this atlas with the unaided eye (using sextants and other measuring devices, but not the magnifying help of a telescope), and is considered by many to be the last great astronomer to have worked in this way. He named his atlas *Firmamentum sobiescianum sive uranographia*, which literally means 'Sobieksi's Heavens, or a Map of the Heavens') after John III Sobieski, the king of Poland at the time, a patron of the arts and sciences, who financially supported Hevelius.

Firmamentum sobiescianum is a significant star atlas for astronomers in part because it includes both more stars and more accurate positions, than were recorded in the *Uranometria* (based on Brahe's measurements). Hevelius detailed these positions in pages of tables, accompanied by 56 beautiful, engraved plates showing all the traditional northern constellations as well as ten new ones (seven of which are still in use). Along with the atlas itself, Hevelius named one of his newly named constellations for the King – Scutum Sobiescianum (today shortened to Scutum) or 'Shield of Sobieski', as well as putting his beloved sextant in the skies and adding several new animals.

He fleshed out the strangely named constellation *Camelopardalis* (literally 'camel leopard' – the Greek word for 'giraffe' – bet you didn't know there was a giraffe in the sky!). *Camelopardalis* does not appear in Ancient Greek charts as it contains only quite faint stars. A single chart also showed the southern skies, based on observations from the English astronomer Edmond Halley (who did use the help of a sextant with 'telescopic sights', but that was allowed). However, Hevelius's atlas showed the constellations backwards (as they would appear on a celestial globe, rather than in the sky), and didn't include labels for the stars, so at times it can be hard to interpret and/or look slightly unusual (for example, Perseus twisting to reach Medusa's head, or Auriga carrying a goat on his back).

Ultimately, this book is a must in our Astronomers' Library, as the last great record of non-telescopic astronomy. It was published in 1690, three years after Hevelius died, by his wife Elisabeth Hevelius, who helped him in his observations and is considered to be one of the first female astronomers.

**Firmamentum sobiescianum
sive uranographia**

From the frontispiece
of the book, this shows
Hevelius presenting his new
constellations to Urania (the
muse of astronomy; centre)
flanked by an array of historical
astronomers. Hevelius is shown
holding items he named in
the constellations (a shield, or
scutum and a sextant) and is
followed by an array of new
animals he placed in the skies.

A star map from the book
showing the new Scutum
Sobiescianum (today just
Scutum) constellation next to
others. The view is a mirror
image of the sky, intended for
use on a celestial globe.

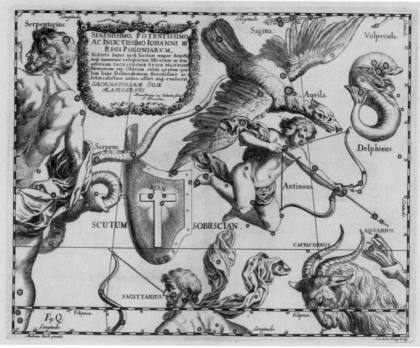

LEFT AND BELOW LEFT
Firmamentum sobiescianum sive uranographia

An all-sky chart from the book, showing stars visible for northern hemisphere viewers.

The main plate for Perseus, but with Auriga and other constellations shown.

Firmamentum sobiescianum sive uranographia

The book's original title page and its accompanying engraving of the author. Technically, *Firmamentum sobiescianum sive uranographia* was the third part of a larger book called *Prodromus astronomie*, which included a preface (*Prodomus*), a catalogue of stars (*Catalogo fixarum*) and the star atlas.

This illustration from the edge of one of the maps in *Firmamentum sobiescianum* (shown on the previous page) shows a cherub holding up a sign saying '*Præstat nudo oculo*' ('the eye is better'), as a response to another cherub offering a telescope. Meanwhile, another cherub observes with a sextant. This is taken to reflect Hevelius's view that telescopic observations of the stars would introduce distortions in positions.

J. HEVELII
PRODROMUS
ASTRONOMIÆ
CUM CATALOGO FIXARUM,
&
FIRMAMENTUM
SOBIESCIANUM.

Atlas coelestis

After the invention of the telescope most astronomers quickly realized how useful this new instrument was for mapping the skies. By the time *Atlas coelestis* (which translates to Heavenly Atlas) by John Flamsteed (1646–1719) was published in 1725, a century after the invention of the telescope, several new astronomical observatories had been built to house these delicate instruments.

This trend included the Royal Greenwich Observatory, which was built in 1675 in what was at the time the outskirts of London, England. The charting of the skies was drawing new interest – and significant funding – as an important method to help with navigation at sea, and particularly to solve the 'longitude problem' (literally, the problem of figuring out what your longitude, or East–West position is at sea). The first appointed 'Astronomer Royal', at the newly built Greenwich Observatory, John Flamsteed was tasked with improving existing charts of the sky with new telescopic observations to better allow the Royal Navy to sail in safety.

Atlas coelestis was the result of Flamsteed's decade-long painstaking efforts to map the skies, and was the first large-scale star atlas based on telescopic observations. The new instrumentation enabled Flamsteed to see fainter stars than are visible with the unaided eye and to measure the locations of these stars very precisely. The book included tables of

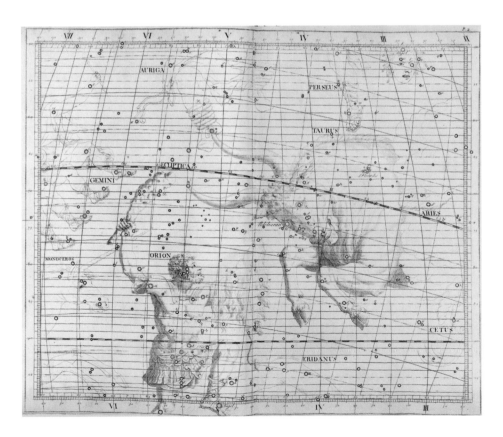

LEFT

Atlas coelestis

Orion, the Hunter, and Taurus, the Bull, in the *Atlas coelestis*.

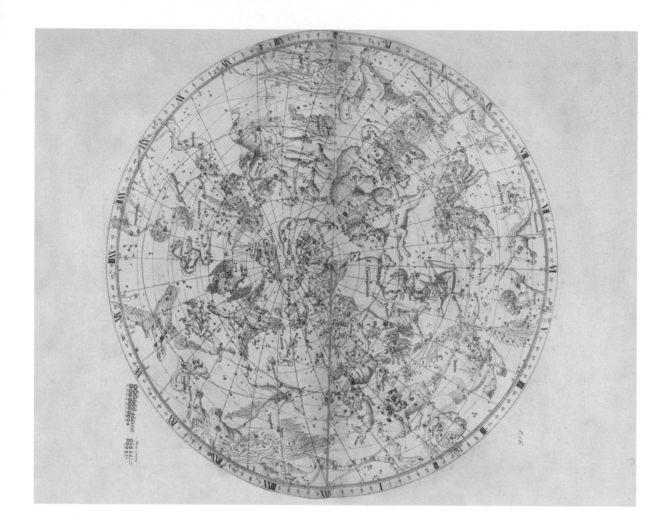

ABOVE

Atlas coelestis

This is plate 26 from *Atlas coelestis,* showing the stars from the northern skies. A key to how the brightness of stars links to the size of the symbol is shown at the lower left.

the coordinates of the stars down to 8th magnitude (about ten times fainter than the faintest stars visible without a telescope) alongside 26 engraved plates with stars on a grid of coordinates overlaid with beautiful artwork of classical Greek constellations.

Most of the star positions in *Atlas Coelestis* were from Flamsteed's observations from Greenwich; however, from this location the southernmost skies are never visible. The English astronomer Edmond Halley (who most notably discovered Halley's Comet) had travelled to the island of St Helena in the South Atlantic in 1677, and he returned with accurate charts of the southern skies. These had previously been used in *Firmamentum Sobiescianum,* which came out about 30 years earlier, but they remained the state-of-the art details of the southern skies (at least in Europe), so Flamsteed also used them for his atlas.

It is an understatement to say that Flamsteed took a really long time to make these maps. Over a decade before the official publication, a pirated copy was released in 1712 by fellow scientists Isaac Newton and Edmond Halley, who had grown so frustrated waiting for the new data, they apparently raided the observatory to steal the information (for the good of science, of course). In the end Flamsteed was such a perfectionist, it wasn't until six years after he died (1719) that his atlas was finally published, around 50 years after it was commissioned. *Atlas Coelestis* (and the accompanying *Catalogus Britannicus*) were his life's work, however as an astronomer he is also noted for the first recorded observation of Uranus (although he didn't know that's what it was), and he observed solar eclipses and comets.

The illustrations of the constellations in the book were done by the English artist James Thornhill (1675–1734), who was commissioned after Flamsteed's death. Thornhill was an English artist, most famous for his large scale mural painting of historical subjects in some of Britain's famous old buildings. He had known Flamsteed and had even painted him, in place of honour on the ceiling of the Great Hall at Greenwich Hospital,

BELOW
Atlas coelestis

The constellation Cetus, the sea monster – or sometimes the whale.

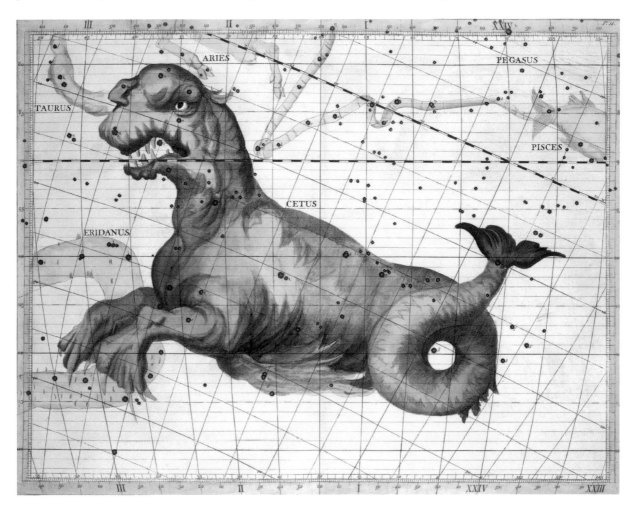

which also featured the great astronomer Tycho Brahe. It also seems Thornhill was left with clear instructions for the constellation art. Flamsteed objected to the 'backwards' representation that had become common (the view of constellations as if looking at them on a celestial globe from the outside) and turned all the images back so they would face the observer – the way the Roman astronomer Ptolemy had described them.

After they waited so long for it to be finished, the atlas became a huge success, and it became the standard reference for astronomers for years to come (although objections were made about its vast size as a book). It wasn't the last of the great star atlases in this style, but it wasn't far off. By the turn of the 19th century, astronomers were becoming more interested in the coordinates of the stars than their representation in constellations. It started to become more common to print star charts with 'stick-figure' constellations as a way to organize the stars into groups, rather than use fanciful illustrations.

BELOW

Atlas coelestis

The constellations Perseus, Cassiopeia and Triangulum.

1. Modern constellation maps

In the modern world, the traditional constellations are mostly of interest for what they say about cultural history than anything to do with the science of astronomy. The stars in a constellation are usually not physically connected – they just happen to be in the same direction from our perspective. However, modern astronomers still use naming conventions tied to (mostly) the traditional Greek constellations – an 'official' set of 88 constellations were adopted by the International Astronomical Union in 1922.

Learning these patterns remains useful for stargazing and amateur astronomy, but when using a professional telescope (and increasingly also some of the telescopes you can buy to use at home), pointing is done by typing coordinates into a computer, so many professional astronomers can identify very few constellations by memory.

Interest is also growing in constellation traditions outside of those adopted by (mostly) European scientists. Most, if not all, cultures have a tradition and stories tied to patterns in the skies. It seems to be something innate to human behaviour the world over that has led us to make dot-to-dot pictures using the skies as a backdrop, as well as to tell such stories to pass down. Many of these star map traditions were forgotten, or ignored, in the rush to map the skies during the period of colonial expansion from Europe. This is perhaps particularly true for the constellations in the southern hemisphere skies, but also the traditions of the indigenous North American peoples and the cultures in Asia and Africa. Our nights are also brightly lit now, and we rarely sit outside to enjoy the darkness, yet we remain fascinated by constellations (most people at least know their 'star sign'!). Relearning traditional stories of the stars provides a way for people to reconnect to their cultural history, and gain better connections to the natural world.

BELOW

John Flamsteed

A portrait of the author of *Atlas coelestis* – and the first Astronomer Royal in England – by the artist Thomas Gibson.

Atlas coelestis

Ursa Major (the Great Bear), with a number of other constellations from the second edition of the book, dating from 1753. This part of the sky contains the Big Dipper or Plough asterism – a collection of stars which forms the body and tail of Ursa Major, and which is both one of the most recognized and most useful to recognize star patterns in the northern hemisphere (since it can be used to find the North star, revealing the directions, and your northern latitude on Earth).

A map of the constellations Boötes, Canes Venatici and the Coma Berenices. From a later, French language, edition, dating from the late 17th century.

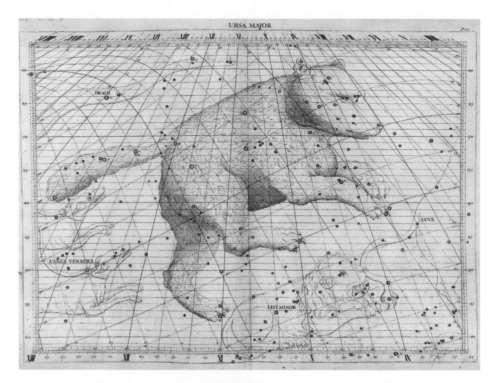

Star mapping ancient and modern

Across the world many indigenous peoples have their own constellation patterns and mythology. This cave art is said to represent the Wintermaker, Moose and Fisher (three constellations of the Ojibwe people, a group of indigenous North Americans).

A modern star chart, showing the constellation of Orion (the Wintermaker to the Ojibwe people) as a stick figure. Orion (and particularly the three stars of its 'belt') is another easy to recognize constellation. Curiously this collection of stars is often turned into some kind of figure in constellation mythology from around the world (the ancient Greeks saw it as a hunter).

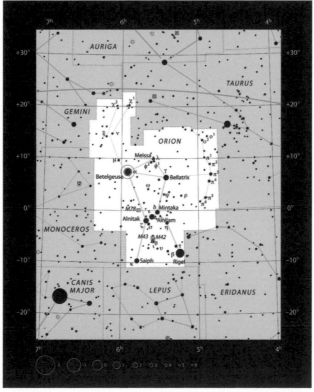

MAPPING
OTHER WORLDS

It is human nature to explore, and the exploration of worlds beyond Earth started as soon as humans began to view the skies as a place that could be visited, rather than a fixed image circling Earth. Before the development of telescopes, the seven 'luminaries' held such a special place that our daily life is set to a seven-day pattern, with the Sun, Moon, Mars, Mercury, Jupiter, Venus, and Saturn listed here in the order of the days of the week named after each of them.

This section of our Astronomers' Library is therefore roughly arranged by these objects. We start with the Sun and the Moon as the brightest, and the most obviously different to the points of light which are stars. After a diversion to discuss eclipses (special astronomical events when either the Moon gets between Earth and the Sun, or Earth gets between the Sun and the Moon), the next section will discuss maps of the planets. For the most part these objects showed no features until astronomers turned telescopes to the skies, and even with a telescope Mercury and Venus are hard to map as 'worlds' so they get little mention except to note the importance of events when they transit the Sun. Maps of Mars, Saturn and Jupiter are more interesting. Astronomers discuss the puzzle of Saturn's changing appearance, caused by the tilt of its rings to the plane of its orbit, and the changing appearance of what we now know are cloud bands on Jupiter. Mars, in particular, has inspired a grand history of detailed observation and mapping, and being the planet most similar to Earth, its changing appearance has caused a lot of speculation about the existence of life (even the possibility of intelligent life) on – or beneath – its surface. Our incomplete collection of books mapping other worlds will conclude with the smallest of the other worlds – comets and how their appearance in the skies has inspired (and/or scared) humankind throughout history.

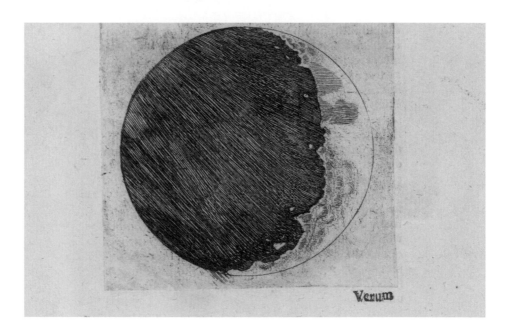

1. Maps of the Moon and Sun

The Moon is our closest celestial neighbour, orbiting Earth at a mean distance of just 240,000 miles (385,000 km) as we travel on our joint path around the Sun. It takes about 29½ days for the Moon to make a complete revolution relative to the position of the Sun, or 27⅓ days relative to the stars (the difference caused by the orbital motion of Earth around the Sun).

For billions of years, the Moon's rotation has been synchronized to its orbit around Earth. This happened quickly after formation, due to a process known as tidal locking. The effect for mapping the Moon is that we only ever see the near side of it from our perspective[1] – the far side (often more commonly referred to as the dark side, even though half of the time it is in full sunlight) is therefore absent in all maps which predate lunar exploration spacecraft.

It has been understood for thousands of years, with evidence from Han dynasty China and from India in 500 CE (see the *Aryabhatiya* on page 119) that the Moon shines because of reflected sunlight; the phases mark its journey around Earth relative to the position of the Sun. Since it is our closest celestial neighbour, it is not surprising that books have contained maps of the Moon for centuries. Notable early drawings include the rabbit in the Moon, as seen by pro-colonial Mesoamericans; East Asian tradition also puts a rabbit in the Moon. In many cultures there are records and drawings of a 'Man in the Moon,' e.g., in the English-language nursery rhyme (of

RIGHT AND OPPOSITE

Moon photography

Humankind's first map of the far side of the Moon, from the Soviet probe Luna 3, which orbited the Moon in 1959.

A 2015 image of the Moon from NASA.

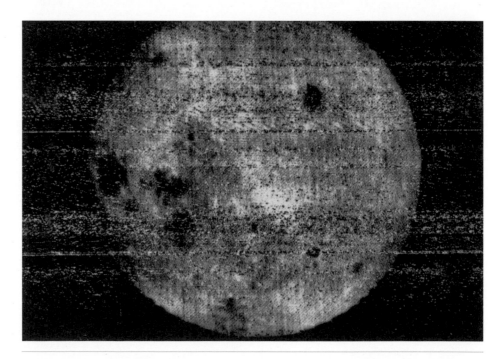

[1]A small wobble termed 'libration' means that averaged over time we see just over half (about 59 per cent) of the Moon's surface.

Drawing of sunspots in Galileo's 1613 work *Letters on Sunspots*.

A detail from *The Florentine Codex* describing pre-Columbian Mesoamerican astronomy (see pages 58–61) showing the 'rabbit in the moon'.

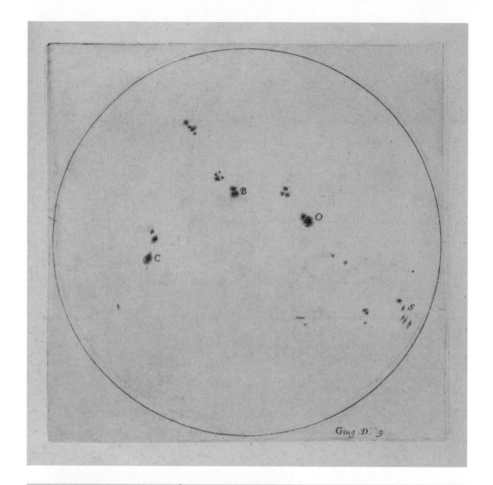

occeppa tepitona
tiubi, occeppa iub
ic iancuican oalm.
ian poliui, poliub
onmiqui, in metzt
cochi: icuein ic co
tiubi, ic tlatbuinac
Aub iniquac uel o.
toa: ommic in me

Old Sun

An ancient record of sunspots, drawn by the English monk John of Worcester in 1128.

which there are many variations) that goes, 'The man in the moon came tumbling down and asked the way to Norwich/He went by the south, and burned his mouth by eating cold plum porridge.' Both of these shapes – the rabbit and the simple face, come from humans putting together understandable patterns in the splotches of light and dark seen on surface of the Moon.

The first telescopic images of the Moon were made by Galileo Galilei in Italy and published in 1610 in his book *Sidereus nuncius* (see page 176). Galileo paid particular attention to features he could see along the line which separates the sunlit Moon from the part in shadow (called the terminator). He used these observations to reason that there were mountains on the Moon in the bright patches we now call the 'lunar highlands'. Following this start, maps of the Moon became more and more detailed, and the far side was first mapped in 1959 when the Russian spacecraft Luna 3 was the first mission to photograph it.

The Sun also holds a special place in humanity's understanding of the cosmos, having been revered by cultures all over the world. Although the Sun's location in the sky is so important to our everyday life, we do not often think about mapping the Sun itself, but the Sun does show visible features, such as sunspots, which persist for a few days at a time. The number of sunspots on the Sun at any one time varies in a roughly 11-year cycle between 'not very many' and 'lots'.[2] The *I Ching*, a book from around 800 BCE in China (often translated as *Book of Changes*), contains the first record of spots on the Sun. The English monk, John of Worcester, published drawings of sunspots in 1128. And again Galileo famously recorded sunspots when he first turned his telescope to the skies, recording them in a series of what he called *Letters on Sunspots* and that he published as an addition to his *Sidereus nuncius* in 1610.

RIGHT AND OPPOSITE

The Man in the Moon

Detail from a 17th-century Venetian manuscript. The symbols around the edge are traditional symbols for the planets, clockwise from top: Jupiter, Saturn, Mars, Mercury, Venus.

The mythical 'man in the Moon' has featured in popular culture for millennia. This image is taken from a story book, circa 1719.

[2] You may find this description unsatisfying. The exact number will depend on the details of your telescope. And I should add here that specialized telescopes or solar filters are needed to view the Sun safely through a telescope (meaning without burning your eyes out), but you might typically see either none at all when the Sun is quiet, up to around ten or more.

Selenographia, sive lunae descriptio

The first really detailed map of the Moon was published in 1647 by the astronomer Johannes Hevelius (1611–1687). Hevelius was arguably the world's last pre-telescopic astronomer, despite having been born after the invention of the telescope and doing some of his significant work using that very instrument. But he is famous for having made the last atlas of stars without the use of a telescope (*Firmamentum sobiescianum sive uranographia* – see pages 41–44) and for his claims that this was the more accurate method to map the stars.

Hevelius enjoyed patronage of a series of Polish kings and it seems like he must have been quite a character. The cherubs in the edges of the illustrations in his books appear to be often making subtle (or not so subtle) comments about his views, or demonstrating the techniques used to make the maps. He built an observatory precariously balanced on the roof of his house, along with an amazing looking 150-foot (45-metre) long telescope which looks as if it was held up by a series of pulleys and winches. His second wife, Elisabeth Koopmann-Hevelius, had been fascinated by astronomy as a child, and was married to the then 52-year-old man when she was just 16, apparently in part because he was an astronomer. She went on to support their astronomical endeavours financially and assisted with observations. For this work she is often noted as one of the earliest female astronomers.

Selenographia, sive Lunae descriptio

Map of the Moon from the book. This shows more than a full disc as Hevelius carefully mapped the libration. Notice the cherubs in the borders demonstrating the techniques of using a telescope, and making precise angular measurements.

A tinted engraving showing Hevelius's 150-foot (45-metre) telescope. The tube is suspended by ropes and pulleys and is not entirely enclosed, so it could only be used in full darkness.

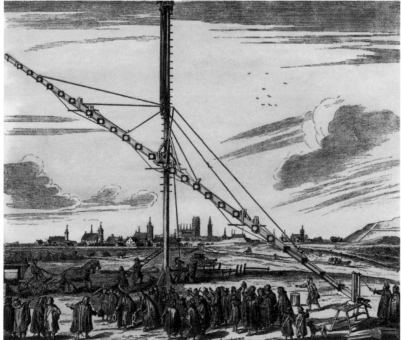

RIGHT
Husband and wife

An image from *Machinae coelestis: pars prior* by Johannes Hevelius, said to show Johannes Hevelius and his wife, Elisabeth, observing the sky with a large brass sextant.

OPPOSITE, LEFT AND RIGHT
Selenographia, sive lunae

The frontispiece from the book, which is full of symbolism and interesting drawings. The figure on the left represents Hasan Ibn al-Haytham (in Latin, 'Alhazen'), a mathematical astronomer from the Islamic golden age shown holding a paper with geometric figures. On the right is Galileo holding a telescope. In the skies above are an image of the Sun, showing sunspots moving across its face, and a map of the Moon. The lower part shows a city scape of Gdansk (Danzig) in Poland, where Hevelius lived and worked.

A series of phases of Venus and Mercury, also in the book.

Fig. M.

This book, *Selenographia, sive lunae descriptio* (which translates to 'Selenography, or A Description of The Moon'), started the field of 'selenography,' the word used to describe studies of the surface of the Moon, and with it the author founded the field of 'lunar topography', by producing the first really accurate map of the Moon (made with the aid of a telescope).

Coming almost 40 years after the first telescopic images, it is not surprising that these maps are much more detailed than the ones from Galileo. Hevelius spent four years on charting the Moon's surface, and during this work he was also the first person to notice the Moon's libration (the wobble which reveals a bit more than 50 per cent of the surface over time). The book contains many detailed drawings of the Moon shown in a variety of phases, along with illustrations of the techniques used to make the maps. It also includes some simple diagrams of the planets Mars and Saturn, Jupiter, the motions of moons around Saturn and Jupiter and the phases of Venus and Mercury.

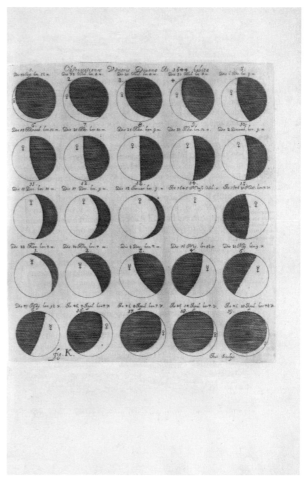

The Sphere of Marcus Manilius Made an English Poem: With Annotations and an Astronomical Appendix

This book contains an annotated English translation of part of an earlier book – the *Astronomica* – along with extensive additional material related to astronomy. The *Astronomica* in its entirety is a five-volume poem from ancient Rome (published in Latin in the 1st century CE) about both astronomy and astrology.

Classical scholars believe the *Astronomica* was written by the Roman poet Marcus Manilius. Not much is known about Manilius, and accounts differ as to even his national origin (he may have been Greek, but at least one historian has claimed he was from Africa). Scholars of the classics also debate the exact date of the book's publication.

The Sphere of Marcus Manilius Made an English Poem was first published in 1675 and was written by the translator and poet Sir Edward Sherburne, an Englishman who fought on the Royalist side during the 17th-century civil war in England. Sherburne translated only the first book of *Astronomica*, the one in which Manilius described a model for the universe (Earth-centric), how the stars and the band of the Milky Way rotate around Earth and gave a discussion of comets as omens of bad luck.

Alongside his translation of the text of the poem, Sherburne included extensive annotations, including comments on his translation choices and notes about astronomical terms and more. This translation takes up the first 68 pages of the book, and his numerous footnotes used all 26 letters in the English alphabet repeatedly. Sherburne then added to this more than 200 pages of appendixes about astronomy, including large format maps showing the entire sky, the Sun and the Moon, diagrams of various 'cosmic' systems (e.g. the Solar System as arranged by Ptolemy, Copernicus and others), and a history of astronomy, including a list of observed comets.

His diagram of the Sun is particularly spectacular, with flames, bands and puffs of smoke, highly reminiscent of the kind of views we get from modern observations today.

BELOW

Sun and Moon

The map of the Sun in the 'Astronomical Appendix' of *The Sphere of Marcus Manilius*. While this seems cartoonish, to a modern astronomer it is impressive how similar this is to modern images of the Sun, hinting at the boiling appearance of convection cells on the surface, sunspots and even solar prominences erupting out.

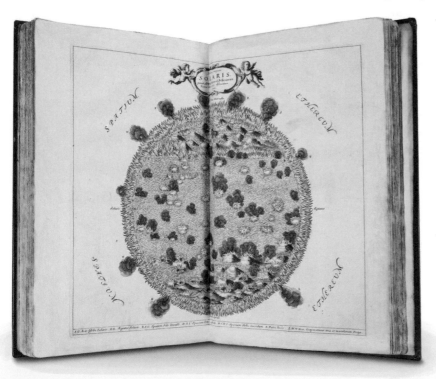

The Moon: Her Motions, Aspect, Scenery and Physical Condition

An early book containing photography of the Moon, this important work was published in 1873 – only 33 years after John William Draper had produced the first ever photograph of the Moon, and still in the early days of any kind of photography.

The author, Richard Proctor (1837–1888), was an English astronomer who was best known for his detailed maps of the planets (especially Mars) and the Moon. He was educated as a lawyer, but he become an incredibly prolific astronomical author, with over 50 books in total. He was able to support his family with the proceeds from his writing, and while some were attempts at rigorous scholarly work, most were written in a style that was more popular among the general public. We have collected just two of Proctor's books in our Astronomers' Library: *The Moon*, which we cover here, and *Other Worlds than Ours* (see page 92). Many others are also worth reading.

The preface of *The Moon* explains that Proctor had intended to take longer over its preparation; however, he was inspired to move more quickly by 'Mr Rutherfurd's magnificent lunar photographs', Lewis Rutherfurd was an American astronomer and

RIGHT
Moon Photography

One of the oldest surviving photographs of the Moon, taken by John William Draper, the English–American scientist, in 1840.

pioneer of astrophotography, who produced many early photographs of the Moon, three of which are published in this book. Proctor also included in the book a 'very full account of the peculiarities of the Moon's motions', together with a complete chapter on the effect of libration (the wobble which means we can see more than half of the Moon from Earth – if we wait long enough). It also contained several chapters about the surface conditions, including comments on the question of whether the Moon might have an atmosphere, work done to search for signs of life on the Moon, and the (then very active) debate over if the common circular features might be impact craters rather than volcanos. Both the illustrations and the content are charming and capture beautifully the understanding of the physical properties of the Moon by 19th-century astronomers.

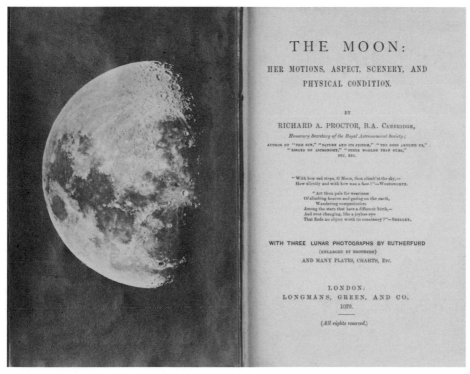

THE MOON:

HER MOTIONS, ASPECT, SCENERY, AND
PHYSICAL CONDITION.

BY

RICHARD A. PROCTOR, B.A. CAMBRIDGE,

Honorary Secretary of the Royal Astronomical Society;

AUTHOR OF "THE SUN," "SATURN AND ITS SYSTEM," "THE ORBS AROUND US,"
"ESSAYS ON ASTRONOMY," "OTHER WORLDS THAN OURS,"
ETC. ETC.

"With how sad steps, O Moon, thou climb'st the sky,—
How silently and with how wan a face !"—WORDSWORTH.

"Art thou pale for weariness
Of climbing heaven and gazing on the earth,
Wandering companionless
Among the stars that have a different birth,—
And ever changing, like a joyless eye
That finds no object worth its constancy ?"—SHELLEY.

WITH THREE LUNAR PHOTOGRAPHS BY RUTHERFURD

(ENLARGED BY BROTHERS)

AND MANY PLATES, CHARTS, ETC.

LONDON:
LONGMANS, GREEN, AND CO.
1873.

(All rights reserved.)

The Moon: Her Motions…

Plate showing how Earth might look from the lunar surface during Earth's daytime (we see South America and some clouds).

Plate demonstrating methods to calculate the distance to the Moon using simultaneous observations from Greenwich and Capetown observatories to create a giant triangle in space (with the base being the North-South distance between the UK and South Africa).

LUNAR LANDSCAPE—"FULL" EARTH.

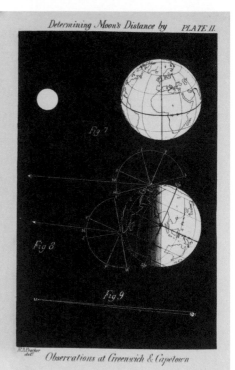

20   THE MOON:

These two stations are not on the same meridian, as will be seen from fig. 7, Plate II., which shows Cape Town more than 18° of longitude east of Greenwich.* At present, however, we shall not take into account the difference of longitude.

Let fig. 8, Plate II., represent a side view of the earth at night, when Greenwich is at the place marked G. Let H *h* be a north and south horizontal line at Greenwich, G Z the vertical, G *p* (parallel to the earth's polar axis) the polar axis of the heavens; and let us suppose that the moon, when crossing the meridian, is seen in the direction G M; then the angle *p* G M is the moon's north polar distance.

Again, let us suppose C to be the Cape Town Observatory, which has at the moment passed from the edge of the disc shown in fig. 8, by nearly 1½ hours' rotation; but let us for the moment neglect this, and suppose the station C to be at the edge of the disc. Let H' C *h'* be the north and south horizontal line at C, C Z' the vertical, C *p'* (parallel to the earth's polar axis) the polar axis of the heavens (directed necessarily towards the south pole); and let us suppose that the moon, when crossing the meridian, is seen in the direction C M'. Then, since the lines G M and C M' are both pointed towards the moon's centre, they are not parallel lines, but meet, when produced, at that point.

Let fig. 9, Plate II., represent this state of things

* This figure is reduced from one of the four summer pictures forming Plate VII. of my "Sun-views of the Earth."

Determining Moon's Distance by PLATE II.

Fig 7

Fig 8

Fig.9

Observations at Greenwich & Capetown

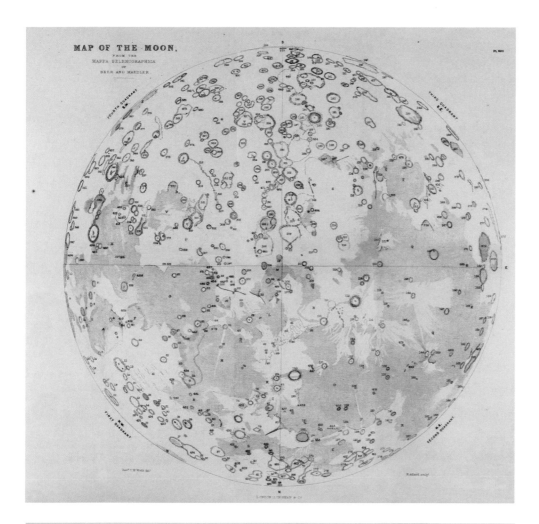

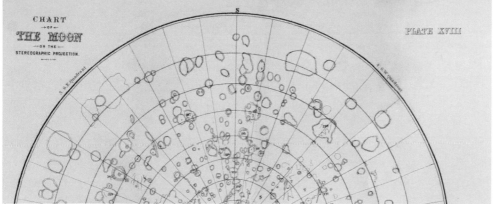

The Moon: Her Motions…

Two of Proctor's maps of the Moon. The top one, 'Map of the Moon from the Mappa Selenographica of Beer and Maedler' shows the Moon as it appears in the sky as mapped by the German astronomers Wilhelm Beer and Johann Heinrich von Maedler and published in three volumes from 1834–1836. On this map, *lunar mare* ('seas', or flat areas) are marked with upper case letters, and craters are numbered. Notable features include G and H (the 'Sea of Tranquillity' and 'Sea of Serenity' respectively; the former where the first Apollo mission landed). Crater 180 is the Tycho Crater (see page 87) and Crater 112 is known as 'Copernicus'.

Lower is a detail from 'Chart of the Moon on the Stereographic Projection', which shows the same map, but presented to flatten the curved lunar surface so that distances between two parts of the map are the same at all points. This means the edges appear stretched out (as we see them at glancing angles from the Earth) relative to the centre.

The Moon: Considered as a Planet, a World and a Satellite

This book was a collaboration between retired engineer James Nasmyth (1808–1890), who had built his own 20-inch (50-centimetre) telescope to observe the Moon, and the professional astronomer James Carpenter (1840–1899), who worked at the Greenwich Observatory. The authors wanted to provide detailed photographs of the surface of the Moon, but techniques did not yet exist for this in the 1870s. So, what they did instead was to observe the Moon by eye and build plaster cast models, which were then photographed to illustrate the book.

The book is sweeping in scope, containing chapters on the overall formation of the Solar System and all the planets in it, discussion of the lunar atmosphere (or lack of it), and how surface features were formed. Nasmyth and Carpenter argued strongly in two separate chapters for the volcanic origin of lunar craters, and many of the diagrams appear to be used to support their view, which time has shown was incorrect (consensus today is that lunar craters are caused by impacts of asteroids and other objects from space). They also included photographs of the back of a man's hand and a shrivelled apple to illustrate the formation of mountain ranges through the shrinking of the interior of the Moon, and an illustration of what a solar eclipse might look like from the surface of the Moon.

RIGHT

The Moon...

By James Nasmyth and James Carpenter, the book was first published in 1874 in London. Two pages from the book: the title page from the 1903 edition in the Library of Congress, New York, and drawings which show a solar eclipse.

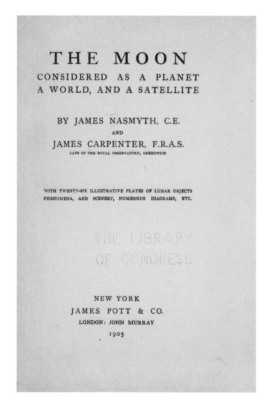

THE MOON

CONSIDERED AS A PLANET
A WORLD, AND A SATELLITE

BY JAMES NASMYTH, C.E.
AND
JAMES CARPENTER, F.R.A.S.
LATE OF THE ROYAL OBSERVATORY, GREENWICH

WITH TWENTY-SIX ILLUSTRATIVE PLATES OF LUNAR OBJECTS
PHENOMENA, AND SCENERY, NUMEROUS DIAGRAMS, ETC.

THE LIBRARY
OF CONGRESS

NEW YORK
JAMES POTT & CO.
LONDON: JOHN MURRAY
1903

Fig. 11.

Fig. 12.

[To face page 68.

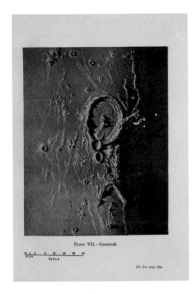

Plate VII.—Gassendi.

SCALE

[To face page 126.]

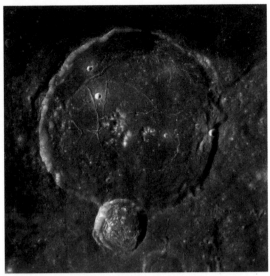

Image from the book, showing a photograph of the crater Gassendi made from a plaster cast based on visual observations.

A modern photograph of the same crater from the Lunar Reconnaissance Orbiter (NASA).

A 'skeleton map of the Moon' from the book, showing names of some features and the locations of major craters (numbered). This was intended to accompany an image, and allow the reader to identify named features.

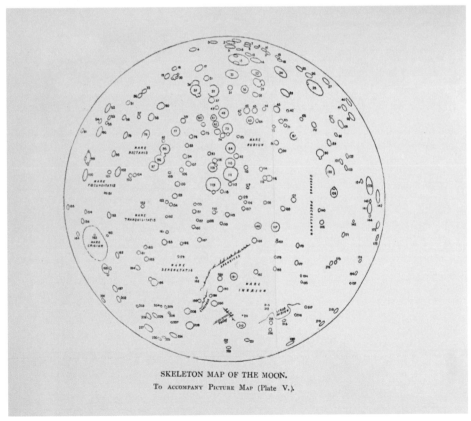

SKELETON MAP OF THE MOON.
To Accompany Picture Map (Plate V.).

The Moon: Considered as…

A rather extraordinary early
photographic image of a hand
(left) and apple (right), used
to demonstrate how mountain
ranges on the Moon might
have formed by the shrinking
of the entire sphere. Also shown
are a photo of the Moon and
drawings of various craters.

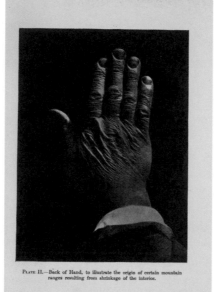

PLATE II.—Back of Hand, to illustrate the origin of certain mountain ranges resulting from shrinkage of the interior.

[To face page 48.

PLATE III.—Shrivelled Apple, to illustrate the origin of certain ranges resulting from shrinkage of the interior of the globe.

[To face page 48.

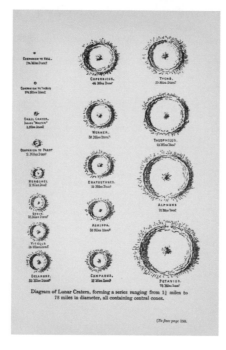

Diagram of Lunar Craters, forming a series ranging from 1½ miles to 78 miles in diameter, all containing central cones.

[To face page 150.

PLATE IV.—Full Moon.

[To face page 56.

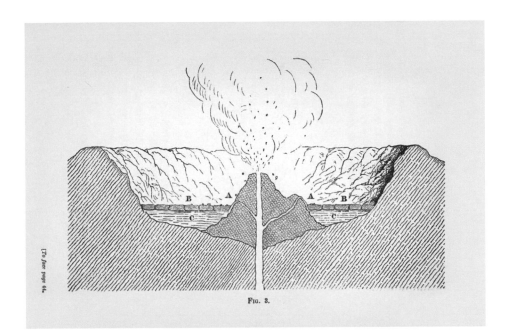

FIG. 3.

PLATE I.—The Crater of Vesuvius in 1864.

The Moon: Considered as...

Two images showing
the now-debunked idea
that craters on the Moon
are volcanic in nature:

(Above): A detailed drawing.

(Below): What appears to be
a photograph of a plaster cast
model including some effect for
outgassing.

Chang'e, the Chinese goddess
of the Moon fleeing to her
home, the Moon.

'Mount Yoshino Midnight
Moon' illustrates the story of a
real princess who, by the light
of a midnight Moon, is said to
have driven away the ghost of
an angry courtier who had been
forced to commit suicide on
Mount Yoshino.

One Hundred Aspects of the Moon: Moon and Smoke

A collection of 100 pieces of art made using the woodblock print method from the Japanese artist Yoshitoshi Tsukioka illustrate this book. Each image shows a famous historical or fictional figure in a scene also featuring the Moon, or in some cases scenes lit by the light of the Moon. While not strictly astronomical in nature, the Moon is such a unifying theme in this lovely series of art that it seems deserving of collection in any astronomer's library. They were first published between 1886 and 1902 in Tokyo by Akiyama Buemon.

The series was printed and released in batches across a period of seven years, and it was so popular at the time that it is said people queued before dawn to obtain their copy whenever new designs were released; they are certainly beautiful.

**One Hundred Aspects
of the Moon**

'Moon and Smoke'.
This woodcut illustrates a
contemporary scene from
late 19th-century Tokyo:
a firefighter tackling a
blaze by moonlight. At
the time, most buildings
in Tokyo were made
of wood and paper,
so firefighters were
particularly heroic figures.

**One Hundred Aspects
of the Moon**

'The Moon of the Moor'
illustrates a moonlit scene in
which a bandit prepares to
attack a flute player. This story
is part of a collection of classical
Japanese stories from around
the 9th or 10th century CE.

**One Hundred Aspects
of the Moon**

The Jade, or Moon Rabbit,
fights with Sun Wukong (the
Monkey King) in this illustration
from *Journey to the West*, a 16th-
century Chinese novel.

2. Lunar and Solar Eclipse

Lunar and solar eclipses have long fascinated humanity, since before books recorded these events. Books containing records of eclipses, while not precisely the same as maps of other worlds, therefore deserve a mention in this part of our library. Some of these explain the geometry of solar and lunar eclipses, while others record the human reactions to these astronomical events.

A solar eclipse occurs when the Moon passes directly between Earth and Sun. By a cosmic coincidence, the Moon appears almost exactly the same angular size in the sky as the much larger, but much more distant Sun. So, when the Moon's orbit lines up just right, the new Moon can completely block the Sun. The alignment needs to be so precise that this blocking happens only for narrow geographical corridors across Earth – called the path of totality. A wider region will see a partial eclipse, where the shadow of the Moon appears to take a bite out of the Sun.

A lunar eclipse is a similar event, except that the Moon, which we see from Earth only by reflected sunlight, falls into Earth's shadow. These eclipses therefore happen at full Moons, and since Earth almost never perfectly shadows the Moon, they are often called 'blood Moons' for the red appearance caused by light from the Sun, which has been reddened by passing through Earth's atmosphere and falling on the Moon.

RIGHT

Compendium in sphaeram

A page from the book showing Earth, the Sun and the Moon in alignment for a solar eclipse.

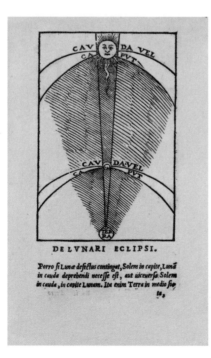

DE LVNARI ECLIPSI.

Porro si Lunæ defectus contingat, Solem in capite, Lunã in cauda deprehendi necesse est, aut uiceuersa Solem in cauda, in capite Lunam. Ita enim Terra in medio sio

Compendium in sphaeram

Compendium in sphaeram is one of many early Renaissance-era commentaries on the book *De sphaera mundi* by Sacrobosco, the scholar and monk. Sacrobosco's famous book (see Chapter 5: Astronomy for Everyone) lays claim to be the most popular book about astronomy ever. *Compendium in sphaeram* by Petrus Velarianus was published in 1537 and contains charming illustrations of the geometry of both solar and lunar eclipses along with other illustrations showing maps of Earth, day-time and night-time on Earth, astronomical coordinate systems and how the phases of the Moon occur.

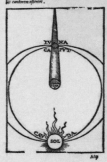

QVID NOX. QVID DIES SIT.

Eꝗ illud ignorandum eſt, Diem aut à tempore
aut à qualitate intelligi, à Tempore quatuor
& viginti horis conſtare, à qualitate eſſe Ae
rem illuminatum à Sole ſupra Horizontem euecto, &
quam diu moram ibi traxerit, ad occaſum vſꝗ. Noctem

vero nihil aliud eſſe quam Terræ vmbram, atꝗ euenire
ſemper vt ſimul ac Sol infra Horizontem deciderit, vm
bra hæc oboriatur, Terramꝗ obfuſcet media ſui parte,
diametro cum Solis progreſſu ſemper mobili, Itaꝗ non
ante Dies eſt, quam Sol iterū ſupra Horizontē emergat.
Accidit

Compendium in sphaeram

Illustrations showing (left)
how night and day happen
at different times in different
places on Earth because of the
shadow of the Sun, and (above)
the geometry of a lunar eclipse.

De eclipsi solari

De eclipsi solari, first published in 1662, contains a satirical poem about the events surrounding a particular solar eclipse which was visible from large parts of Europe on 12 August 1654. The author, Jakob Balde (1604–1688), was a German native, and famous for his poetry in classical Latin.

The book starts with several whimsically illustrated pages (see left and right). One shows a bearded man gazing up at the covered Moon. Some readers have speculated this is intended to represent Galileo, who died just a few years earlier. Certainly an object which appears similar to Galileo's famous telescope is clearly visible among the scientific objects around the man. There are then two pages which show the extent of the Moon's shadow in different European cities. The bulk of the text contrasts the explanation of the eclipse event by a mathematician and a poet. They have contrasting views of the event as being the result of either natural causes or being a divine intervention meant to signal some omen against humanity's sins, which Balde used to capture the general sense of the significance of an eclipse at this time.

While the scientific world in Europe was excited about the 1654 eclipse event, there was a lot of anxiety around it among the general public at the time. Pamphlets were even published to reassure people that the eclipse was not thought to be a prediction of any particular disaster.

Curiously, the eclipse may have actually impacted events at a European battle – at the time this eclipse occurred, the armies of Russia and Poland were at war – and the eclipse event occurred during the Battle of Szklow on 12 August. Russian forces were so surprised by the eclipse of the Sun that the much smaller Polish army was able to claim victory (although who actually won is still somewhat debated).

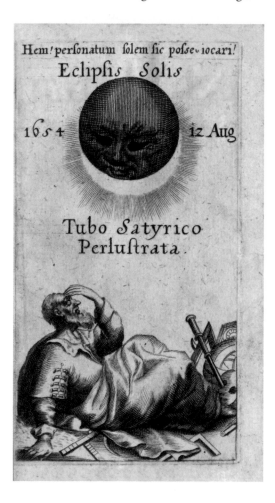

BELOW

De eclipsi solari

The central words read 'Our world is duped by this shadowy masked Moon'. This illustrates the expected extent of the eclipse as seen from various European cities, as named.

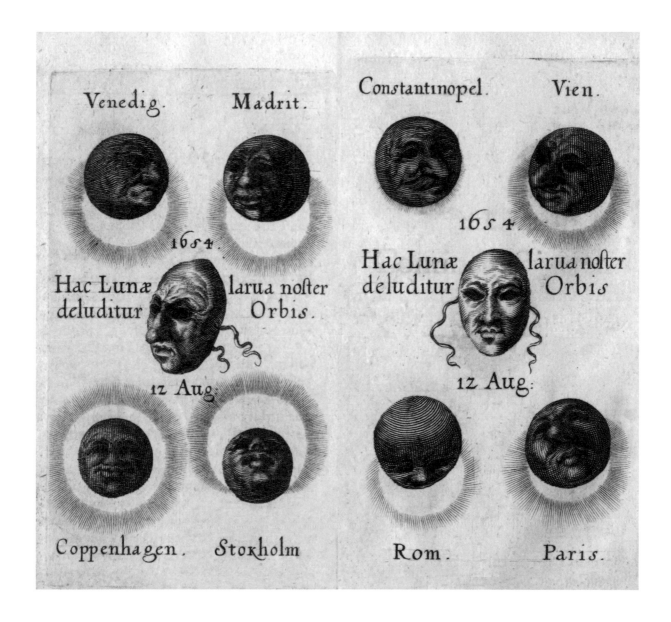

Canon der finsternisse

The Canon of Eclipses is a compilation of the details (timing and path of totality) of over 13,000 eclipses. This includes every known lunar or solar eclipse since 1207 BCE, as well as predictions for eclipses going forward to 2161.

While initially published in 1887, the book was reprinted as recently as 1962; the foreword of that edition notes how useful the book had been to both astronomers and historians, and that as late as 1962 – 75 years after its first publication – it remained the most complete set of calculations of all lunar and solar eclipses available.

The author, Theodor Ritter von Oppolzer (1841–1886), was a very mathematical astronomer. He was born in Prague, and his father, a famous pathologist, insisted he study medicine. However, von Oppolzer had a passion for astronomy since childhood. As soon as he was able, he built an observatory at his own expense, and he was so talented at the subject that he became a professor of theoretical astronomy in Vienna and wrote articles on the standardization of the length of the day. He had a particular talent for making complex calculations about the orbital motions of the Moon, Earth and other planets. Information about the orbit of both Earth and the Moon are what are needed to make accurate predictions of the timing and geography of eclipses. Von Oppolzer's Canon of Eclipses compiled these calculations for thousands of eclipse events from the past and into the future (from 1207 BCE to 2161 CE). He was aided in compiling the calculations by a team of astronomers. The book itself is mostly hundreds of pages of tables showing parameters which can be used to find the timing and geography of an eclipse. At the end of the book are 148 diagrams showing the paths of eclipse totality across maps of Earth.

Today's equivalent to Theodor Ritter von Oppolzer and his *Canon* is undoubtedly 'Mr Eclipse', Fred Espenak, who worked for NASA to create predictions for five millennia of eclipses (11,898 solar eclipses occurring between 1999 BCE and 3000 CE). His personal website (MrEclipse.com) and series of books of tables of eclipse, including custom maps for contemporary events (e.g. *Road Atlas for the Total Solar Eclipse of 2024*) provide detailed information for today's 'eclipse chasers'.

Culturally, we no longer believe eclipses predict disasters or are otherwise bad omens, but experiencing a total solar eclipse is an astronomical experience like no other. It is well worth a trip to get into the path of totality at least once in your life if you can. You will experience a natural event which leaves you with a deeper understanding of your life on a planet in space, and experience for yourself the sense of awe that presumably led early humans all over the world to either fear or worship these special moments in our planet's motion through space.

RIGHT AND OPPOSITE
Canon der finsternisse

The title page of the original edition, published in Vienna.

Map showing the paths of totality for several eclipses detailed in the tables.

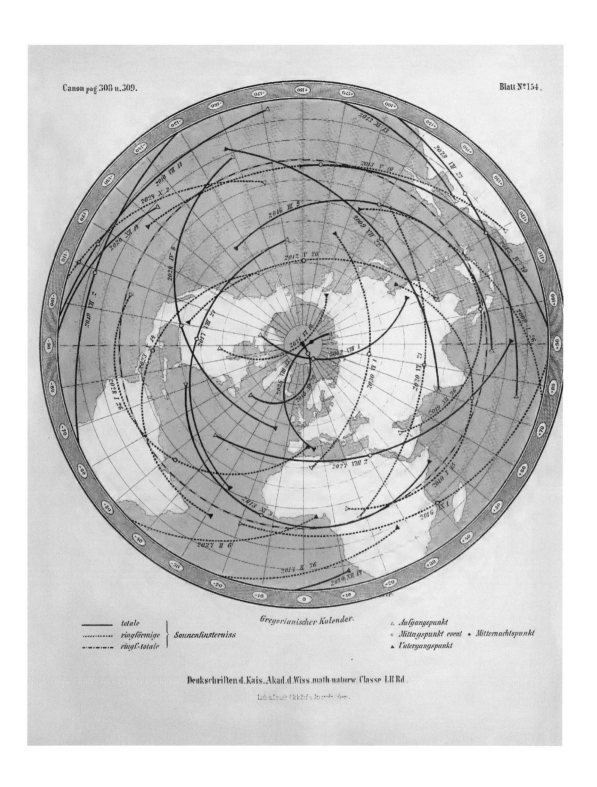

Gregorianischer Kalender.

———	*totale*		△ *Aufgangspunkt*
··········	*ringförmige*	} *Sonnenfinsterniss*	○ *Mittagspunkt event* • *Mitternachtspunkt*
– – – –	*ringf.-totale*		▲ *Untergangspunkt*

Denkschriften d.Kais..Akad.d.Wiss.math-naturw.Classe LII.Bd.

Lith.u.Druck d.kk.Hof u.Staatsdr. Wien.

RIGHT

Canon der finsternisse

The eclipse listings as seen in the book. There are hundreds of pages of this information.

Nr.	μ	γ	log n	G	K	log sin g	log sin k	log cos g	log cos k	log sin δ'	log cos δ'	N'	bei ⊙ Aufgang λ	φ	im Mittag λ	φ	bei ⊙ Untergang λ	φ	F'
7651	333°83	+0·8340	9·7528	141°31	96°70	9·6612	9·9733	9·9489	9n5321	9·4868	9·9786	111°0	−103	+68	+35	+81	+114	+34	t*
7652	296·59	−0·2856	9·7073	319·06	96·72	9·6607	9·9753	9·9489	9·5154	9n5048	9·9766	69·8	−12	−35	+67	−36	+124	+3	r
7653	217·16	+0·0734	9·7665	132·60	96·29	9·6516	9·9816	9·9513	9n4544	9·5395	9·9723	107·7	+71	+20	+143	+25	−158	−13	t*
7654	284·47	+0·3971	9·7030	308·82	95·97	9·6477	9·9849	9·9523	9·4147	9n5568	9·9698	73·8	+15	+7	+72	+3	+122	+36	r*
7655	112·04	−0·6757	9·7589	123·13	95·25	9·6387	9·9892	9·9545	9n3434	9·5742	9·9671	103·8	−171	−25	−116	−22	−71	−50	t
7656	311·58	+1·0595	9·7221	297·76	94·50	9·6315	9·9925	9·9561	9·2645	9n5872	9·9649	78·5	—	—	—	—	—	—	p
7657	139·54	+1·2165	9·7215	83·59	89·09	9·5777	9·9997	9·9665	8·5923	9·5754	9·9669	87·6	—	—	—	—	—	—	p
7658	308·69	−1·4890	9·7351	112·71	93·69	9·6232	9·9951	9·9578	9n1735	9·5940	9·9636	99·3	—	—	—	—	—	—	p
7659	278·34	−1·0570	9·7602	254·05	87·96	9·5628	9·9981	9·9689	8n9728	9n5480	9·9711	95·8	—	—	—	—	—	—	p*
7660	179·11	+0·4862	9·7042	72·13	87·79	9·5572	9·9976	9·9697	9·0167	9·5385	9·9724	83·6	+109	+21	+179	+49	−101	+33	t*
7661	156·74	+0·3756	9·7649	242·41	87·07	9·5425	9·9949	9·9719	9n1858	9·4957	9·9776	99·3	+133	−12	−158	−41	−80	−30	t
7662	187·58	−0·2661	9·7103	59·85	86·96	9·5372	9·9940	9·9726	9·2174	9·4807	9·9792	80·0	+119	−24	+174	+2	−127	−5	r*
7663	15·82	+0·3238	9·7481	230·05	86·70	9·5238	9·9904	9·9742	9n3159	9n4187	9·9845	102·4	−71	+30	−14	+4	+47	+6	r
7664	271·59	−0·9962	9·7343	47·17	86·70	9·5205	9·9894	9·9747	9·3388	9·3971	9·9860	77·0	+120	(−73	—	—	+138	−67	r
7665	150·07	+1·0875	9·7214	217·04	86·93	9·5090	9·9856	9·9760	9n4021	9n3039	9·9910	104·9	—	—	—	—	—	—	p
7666	324·78	+0·9480	9·7648	359·33	90·07	9·4872	9·9785	9·9785	9·4872	7n5738	0·0000	72·1	46	+54	+27	+85	+100	+90	t
7667	284·59	−1·1040	9·7037	167·83	91·22	9·4874	9·9796	9·9785	9n4265	8·8317	9·9990	107·5	—	—	—	—	—	—	r*
7668	206·71	+0·2641	9·7609	346·28	91·37	9·4873	9·9799	9·9785	9n4735	8n8823	9·9987	72·6	+88	−2	+151	+12	−145	+33	t*
7669	316·95	−0·3365	9·7205	154·51	92·36	9·4949	9·9824	9·9777	9n4461	9·1466	9·9957	106·4	−20	−3	+40	−12	+100	−36	r
7670	40·20	−0·4547	9·7382	333·16	92·47	9·4969	9·9827	9·9775	9·4426	9n1692	9·9952	73·7	−114	−43	−37	−37	+27	−11	r
7671	95·66	+0·4331	9·7473	141·91	93·08	9·5081	9·9861	9·9761	9n3947	9·3128	9·9906	104·7	−171	+39	−93	+38	−28	+11	t*
7672	129·43	−1·2077	9·7119	320·02	93·18	9·5121	9·9867	9·9756	9·3861	9n3340	9·9296	75·6	—	—	—	—	—	—	p
7673	224·01	−1·3577	9·7666	98·73	91·22	9·5748	9·9994	9·9670	8n7249	9·5705	9·9677	93·3	—	—	—	—	—	—	p
7674	325·11	+1·1440	9·7650	130·23	93·33	9·5258	9·9903	9·9741	9n3205	9·4190	9·9845	102·7	—	—	—	—	—	—	p
7675	203·91	+1·1462	9·7066	272·47	90·36	9·5839	9·9999	9·9654	8n1833	9n5836	9·9655	89·1	—	—	—	—	—	—	p
7676	109·66	−0·6504	9·7534	87·28	89·58	9·5924	9·9999	9·9639	8·2320	9·5920	9·0640	88·9	−160	−38	−109	−18	−58	−36	t
7677	259·35	+0·4173	9·7301	260·27	88·45	9·6037	9·9992	9·9618	8n7944	9n5984	9·9628	93·9	+48	+26	+101	+1	+157	+19	r*
7678	279·54	+1·1169	9·7272	76·32	87·79	9·6100	9·9983	9·9606	8·9465	9·5995	9·9626	84·5	+18	+1	+80	+30	+147	+11	r*
7679	64·55	0·2901	9·7556	248·94	86·57	9·6215	9·9959	9·9583	9n1401	9n5964	9·9632	98·6	−133	8	66	−40	+11	−23	r
7680	340·65	+0·9114	9·7065	65·64	86·03	9·6264	9·9943	9·9572	9·2057	9n5926	9·9639	79·9	−88	+49	165	(+89	+154	+63	r*
7681	295·63	−0·9495	9·7657	238·56	84·98	9·6362	9·9903	9·9549	9n3198	9n5787	9·9663	103·0	−49	−52	−121	(−81	−137	−67	t
7682	130·84	1·1953	9·7157	30·35	83·51	9·6697	9·9642	9·9464	9·5907	9·4119	9·9850	66·2	—	—	—	—	—	—	p
7683	349·02	+1·0740	9·7410	203·58	84·79	9·6762	9·9567	9·9446	9·6285	9n3238	9·9901	115·8	—	—	—	—	—	—	p
7684	244·37	0·3992	9·7419	21·87	84·43	9·6780	9·9548	9·9440	9·6369	9·2964	9·9913	63·8	+64	−48	+121	15	179	+3	t
7685	93·39	+0·3792	9·7147	195·24	85·69	9·6836	9·9481	9·9424	9·6635	9n1564	9·9955	117·7	147	+49	−88	+17	−29	−5	r*
7686	93·95	+0·3395	9·7627	14·02	86·00	9·6831	9·9475	9·9426	9·6662	9·1209	9·9962	62·1	−159	−8	−99	+30	−20	+47	t*
7687	104·08	0·3471	9·7029	187·18	87·86	9·6872	9·9426	9·9412	9n6827	8n8420	9·9986	118·9	−166	+9	−110	−27	−37	−49	r
7688	340·83	+1·0380	9·7640	6·49	88·05	9·6840	9·9435	9·9424	9·6804	8·7942	9·9992	61·3	—	—	—	—	—	—	p
7689	117·07	−1·0607	9·7149	179·34	90·20	9·6867	9·9415	9·9415	9n6867	7·8038	0·0000	119·1	—	—	—	—	—	—	p
7690	359·51	−0·9772	9·7301	337·10	95·73	9·6774	9·9559	9·9443	9·6323	9n3136	9·9906	64·0	+136	(−72	—	—	+99	−50	r
7691	85·12	+0·9016	9·7540	150·02	96·46	9·6699	9·9639	9·9464	9n5928	9·4074	9·9853	113·9	+113	+75	(+105	(+85	+6	+39	t*
7692	56·30	−0·2989	9·7065	328·38	96·66	9·6710	9·9852	9·9461	9·5850	9n4282	9·9838	66·5	−131	30	52	−35	+4	+6	r
7693	330·01	+0·1455	9·7666	141·81	96·74	9·6628	9·9726	9·9484	9n5369	9·4843	9·9788	111·2	44	+28	+31	+27	+91	−12	t*
7694	43·84	+0·3861	9·7030	318·79	96·71	9·6603	9·9756	9·9490	9·5132	9n5055	9·9764	69·9	−106	+2	−48	+6	+2	+40	r*
7695	222·29	−0·6029	9·7577	132·87	96·30	9·6515	9·9814	9·9513	9n4567	9·5377	9·9725	107·8	+76	−18	+133	−19	−179	−50	t
7696	75·72	+1·0507	9·7235	308·42	95·90	9·6463	9·9852	9·9526	9·4093	9n5574	9·9697	74·0	—	—	—	—	—	—	p
7697	241·39	+1·2975	9·7199	94·86	90·75	9·5961	9·9908	9·9633	8·4874	9·5947	9·9635	91·9	—	—	—	—	—	—	p
7698	52·79	−1·4260	9·7332	123·03	95·21	9·6375	9·9893	9·9547	9n3412	9·5735	9·9672	103·7	—	—	—	—	—	—	p
7699	47·87	−1·0652	9·7608	266·34	89·46	9·5826	9·9999	9·9657	8n3538	9n5818	9·9658	91·4	—	—	—	—	—	—	p
7700	277·47	+0·5669	9·7043	83·72	89·11	9·5769	9·9997	9·9666	8·5826	9·5747	9·9670	87·6	+4	+30	+82	+57	+164	+34	r*

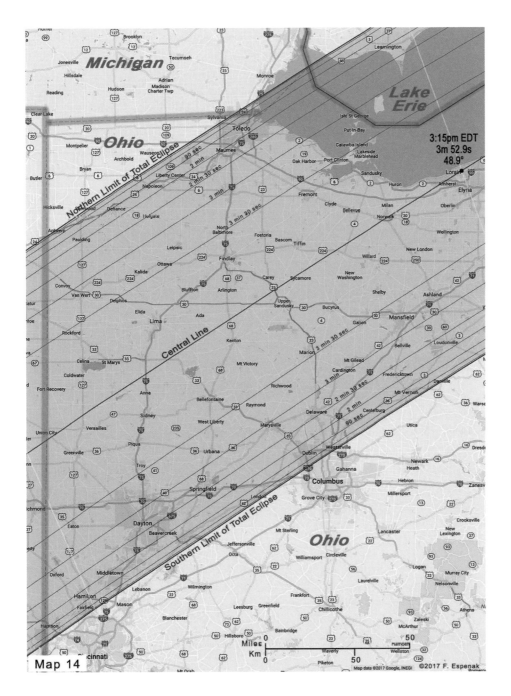

Map 14

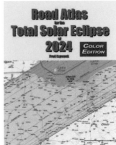

ABOVE AND RIGHT

Road Atlas for the Total Solar Eclipse of 2024

Fred Espenak – 'Mr Eclipse' – has listed events that will take place up to 3000 CE. This is the cover and a sample map from his book about the 2024 solar eclipse in North America.

OPPOSITE, CLOCKWISE FROM LEFT
Images of Saturn

An image of Jupiter taken by
the Hubble Space Telescope in
2020. The moon Europa is also
visible to the left of the planet.

Painting of astronomers
observing Jupiter, by the Italian
artist Donato Creti in 1711.
The planet and three of its
moons are shown as they would
appear in a telescope.

Jupiter (upper in two views)
and Mars (lower) sketched
by Huygens and published in
his *Systema saturnium* (1659).
The cloud bands on Jupiter are
apparent, as well as plausibly
the pole caps on Mars.

3. Mapping the Planets

For much of human history the five bright planets, or 'wandering stars' (Mercury, Venus, Mars, Jupiter and Saturn) were only distinguishable from stars by their strange motions through the zodiacal constellations. Those with exceptionally keen eyes are said to be able to make out the crescent phase of the planet Venus when it is closest to us, and some observant people might also have noticed how the planets do not twinkle like the stars as they set… but it was not until the telescope was turned to the skies that planets began to be seen as other worlds to be mapped.

Galileo started this process when he noted in 1610 that through his telescope, the planets appear as small discs, rather than points of light like the fixed stars. Most of his early reports of these telescopic findings were descriptive texts, but later he made a series of drawings of the phases of Venus, the discs of Saturn, Jupiter, and Mars and the 'ears' on Saturn.

Jupiter and Saturn: The two largest planets in the Solar System, Jupiter and Saturn at times appear as bright (or sometimes brighter) than their more nearby siblings. They move slowly through the zodiac constellations: Jupiter taking 12 years and Saturn 29 years. Jupiter was famously one of the first telescopic targets for Galileo (see page 179) who tracked moons orbiting the planet. One of the planet's most famous features is its Great Red Spot, a storm larger than Earth itself, which may have been noted as early as 1664, and in recent years it appears to be weakening. Saturn is most noted for its beautiful rings. It is also the planet with the largest collection of moons in the Solar System, although Jupiter competes for this title at times as more and more small moons are discovered around both planets.

Systema saturnium

The *Systema Saturnium* (Systems of Saturn) was published roughly 50 years after Galileo's first telescopic observations. In it, Christiaan Huygens (1629–1695) proposed for the first time that Saturn's strange and changing appearance through a telescope was caused by it having 'a thin, flat ring' around the main body of the planet, that was 'inclined to the ecliptic' – meaning that from our viewpoint on Earth we see it from different angles as Saturn moves around the Sun.

Huygens was a Dutch scientist during the scientific revolution in Europe. He was born in The Hague into a prominent and well-connected Dutch family, many of whom worked as diplomats or with the Dutch royal family. Huygens instead pursued science, where he made significant contributions to the fields of mathematics, optics and mechanics, as well as astronomy. Within the field of astronomy he is best known for his work on the planet Saturn however, he also is said to have been the first person to draw a map of Mars with any features.

Systema saturnium also contained sketches of both Jupiter and Mars, as well as the first known sketch through a telescope of the Orion nebula (a star-forming region in our galaxy, which is barely visible to the unaided eye in the 'sword' of Orion.

Huygens also used the book to support the Copernican model of the Solar System (see Chapter 4: Developing Our Model of the Universe), estimating the relative distances between the planets and the Sun by using their angular diameters, along with an estimate of the distance from the Sun to Earth, to put a scale to the Solar System. As part of this, in *Systema Saturnium* he discussed how he used telescopic observations to measure the angular sizes of the planets by using thin rods inserted into the telescope (to set a scale). In this way he was perhaps the first person to use a telescope to make scientific measurements rather than just observe the night skies, and so his book is an essential part of any astronomer's library.

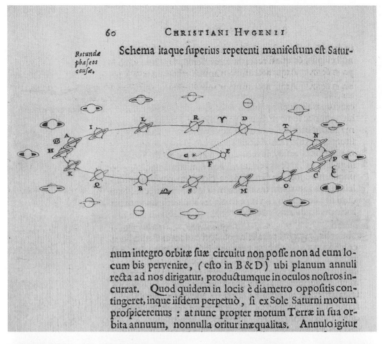

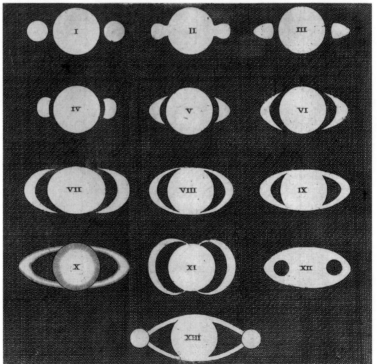

Saturn in photographs

Modern photographic
imagery of Saturn (taken by
the Hubble Space Telescope),
showing how different the
planet can appear throughout
the Saturnian year. Compare
this to Huygens' diagram,
opposite.

Huygens' estimates of the
relative sizes of the planets and
the Sun (Sol.) in his later book,
Cosmothereous (1698).

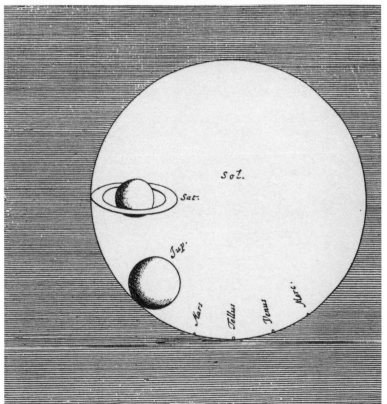

Other Worlds Than Ours

The second of two books we are collecting from the prolific 19th-century English astronomical author Richard Proctor (see page 68) is *Other Worlds Than Ours*. It carries the rather wordy subtitle 'The Plurality of Worlds Studied Under the Light of Recent Scientific Researches' and as a record of the early 19th century understanding of the cosmos, it is fascinating.

The chapters of the book sweep across the entirety of the known universe at the time from Earth to the Sun, and all the planets, including mention of 'Uranus and Neptune, the Arctic planets', then moons (including ours), meteors and comets. Towards the end are two chapters which cover matters to do with 'other Suns', including thoughts on their distribution in shape (i.e. a map of our galaxy) and one on 'The Nebulae: are they External Galaxies?', a question which at the time this was written – roughly 50 years before Edwin Hubble would publish an estimated distance to the Andromeda galaxy – was very much not settled. Proctor ended his book with a chapter entitled 'Supervision and Control' in which he lays out his viewpoint on how religion can be reconciled with an ever-widening understanding of worlds above Earth, which Proctor assumes are 'peopled with millions of living creatures', perhaps even intelligent ones.

The book is perhaps most famous for the publication of an early map of Mars showing several recognizable features. Proctor drew the map based on detailed drawings from William Rutter Dawes (1799–1898). Dawes (whose father, also called William

Dawes, was also an astronomer) observed the planet from his home observatory during its 1864 opposition and made many charts. 'Opposition' refers to the point in the planet's orbit when it is directly opposite the Sun in Earth's skies. This point brings it closest to Earth, and therefore it is largest in angular scale and optimal for detailed observation from the ground. In his map based on Dawes's drawings, Proctor named several features, including two after Dawes. However, both the map and these names were soon superseded and are not used today.

The book also includes a drawing of the planet Jupiter by the astronomer John Browning. Browning is mostly known for his precise instrumentation, and the comments note that Jupiter was one of his own reflectors, indicates an appearance not uncommonly seen, a dark streak extending obliquely across the planet's equatorial regions. The number of belts is singularly variable. Sometimes only one has been seen, at others there have been as many as five or six on each side of the planet's equator. In

Fig. 1.—The Planet Jupiter (Browning).

the course of a single hour, Cassini saw a complete new belt form on the planet, and on December 13, 1690, two well-marked belts vanished completely, while a third had almost disappeared in the same short interval of time.

But if we seem to recognize here the action of forces much more intense than those which influence the condition of the earth's atmosphere, we have still

minuter peculiarities of structure, we are led to the conclusion that the Milky Way, judged according to the fundamental hypothesis of Sir W. Herschel, has some such shape as I have endeavored to exhibit in the accompanying figure. Although I have not indicated here the corrugations of the ring, nor a tithe of the various overlapping layers which would be required to account for the appearance of the Milky Way between Centaurus and Ophiuchus, yet the deduced figure is by no means inviting in its

Fig. 4.—The Galactic Flat Ring, modified in accordance with the observed peculiarities of the Milky Way.

simplicity. It is, however, absolutely certain that the sidereal system, as far as its more densely aggregated star-regions are concerned, has some such figure as this, if we are to accept the principle of Sir W. Herschel's star-gaugings.

Now, in turning our thoughts to the recognition of a more simple explanation of observed appearances, it will be well that we should consider some peculiarities of the Milky Way which we have not yet attended to. In the first place, I would invite attention to a peculiarity observed by Sir John Her-

observed 'with one of his own reflectors'. Proctor uses the map to note on the variability of the bands on Jupiter (which we now know are cloud bands) comparing this drawing to observations by Italian astronomer Giovanni Cassini (1625–1712).

Moving beyond our Solar System, Proctor also provides diagrams illustrating what was understood at the time about the distribution of stars in space around the Sun (aka our Galaxy). He describes our Galaxy as a disc, with the Sun at the centre, similar to the shape which had been suggested by William Herschel (aided by Caroline Herschel) almost a century earlier. The shape he draws, which looks a bit like a pancake, is in fact about right, although today we know the Sun is not in the centre, but about two-thirds of the way out into the disc. This section also contains speculation about the possible spiral structure of the Milky Way and how that would look from our viewpoint.

'A Chart of Mars' from an 1896 edition of the work. The names for features suggested by Proctor were not adopted generally, so we use different ones today. The chart was based on drawings by the astronomer William Rutter Dawes.

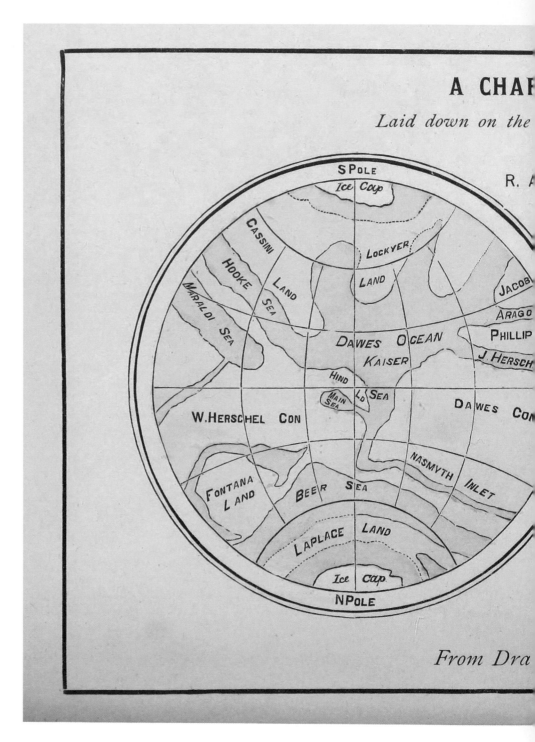

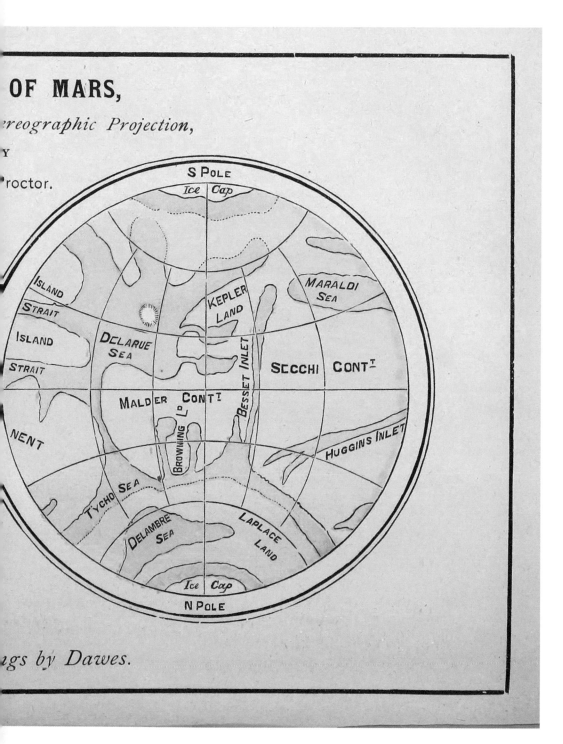

OF MARS,

...reographic Projection,

...Y

...roctor.

S POLE
Ice Cap
ISLAND
STRAIT
ISLAND
STRAIT
DELARUE SEA
KEPLER LAND
MARALDI SEA
MALDER CONT?
BROWNING L?
BESSET INLET
SECCHI CONT?
...NENT
HUGGINS INLET
TYCHO SEA
DELAMBRE SEA
LAPLACE LAND
Ice Cap
N POLE

...gs by Dawes.

Mars

Mars holds a special place in the interest of humans, being the most similar of all the other Solar System planets to Earth. Christian Huygens (see page 88) drew the first map of Mars, and another early map by Richard Proctor (see pages 94–95) was based on drawings by William Rutter Dawes made in 1864. While ideas about life on all the other planets abound in some of the books we have collected, Mars in particular has inspired astronomers as a possible extraterrestrial world. It is a challenging object to map from the Earth, varying in scale from something just barely big enough to be resolved by a ground-based telescope, to a mappable world at opposition (when it and the Earth are on the same side of the Sun).

La vita sul pianeta Marte

A particularly notable example in the history of astronomers viewing Mars as a world teeming with life is *La vita sul pianeta Marte* (Life on Mars) by the Italian astronomer Giovanni Schiaparelli (1835–1910). In this book, Schiaparelli published his detailed map of Mars based on observations during the opposition of 1877. In these observations he named what he saw as the 'seas' and 'continents' of Mars, and described a network of *canali* ('channels'), which he believed were a kind of natural river network on the planet's surface. In the English language world this description was mistranslated as 'canals', and these observations were interpreted as revealing an intelligent civilization living on the planet, an idea which took off in the imagination of people all around the world. By the early part of the 20th century, however, most astronomers had concluded that Schiaparelli's *canali* were not even real as natural features, but rather likely an optical distortion. However, the idea of intelligent life on Mars has remained a popular topic for science fiction writers ever since.

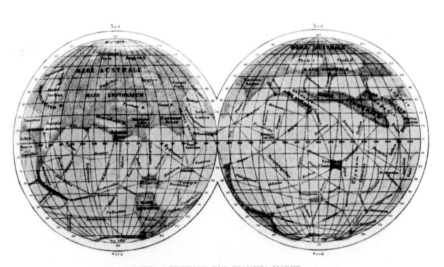

CARTA GENERALE DEL PIANETA MARTE
secondo le osservazioni fatte a Milano dal 1877 al presente.

NB. - Le linee o strisce oscure che solcano i continenti sono in questa carta presentate nel loro stato semplice cioè come appaiono quando non sono geminate.

Mars and Its Canals

One astronomer who was particularly inspired by Schiaparelli's Martian 'canals' was Percival Lowell (1855–1916). Lowell was a successful American businessman and diplomat who had travelled extensively in East Asia, particularly Korea and Japan. After moving back to the USA in 1893, he became fascinated by Mars, and he used some of his wealth to found an observatory (the still-existing Lowell Observatory in Flagstaff, Arizona – perhaps most famous as the location from which Pluto was discovered). Lowell spent 15 years on an extensive study of Mars through his telescopes, making many drawings, which he published in a series of three books: *Mars* (1895), *Mars and Its Canals* (1906), and *Mars as the Abode of Life* (1908). We'll collect *Mars and Its Canals* here, as an example of these and the role they played in popularizing the idea of canals on Mars as evidence (now thoroughly debunked) of intelligent life on Mars.

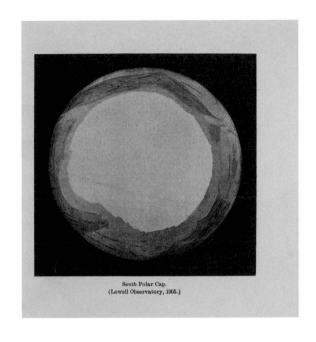

South Polar Cap.
(Lowell Observatory, 1905.)

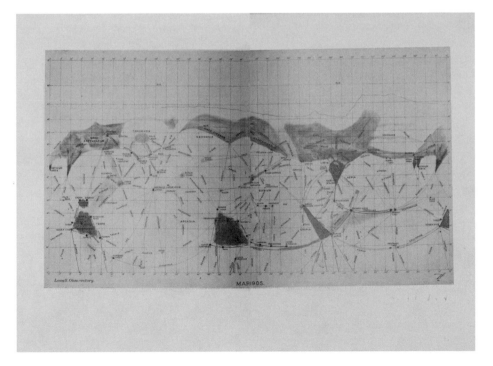

ABOVE AND LEFT

Mars and Its Canals

Image of Mars' 'South Polar Cap' from Lowell's book *Mars and Its Canals*.

Map of Mars by Percival Lowell.

The Inferior Planets

The inner two planets, Mercury and Venus, which are sometimes known as 'the inferior planets' are challenging to map in detail from Earth. Mercury is, for good reason, neglected by telescopic observers, as it is always close to the Sun in the sky – either hidden in the Sun's glare or appearing only close to the horizon around dawn and dusk. Meanwhile Venus, while famous for its phases (which Mercury also has) is almost completely featureless through an optical telescope. We now know the reason is because it is covered by a thick layer of clouds, which completely obscure any features on the surface of Venus.

However, one special feature of both inner planets is that they both can – and do – at regular intervals, pass directly between Earth and the Sun. These events are known as transits, and when they happen, either planet appears as a small dot, or shadow, which moves slowly across the face of the Sun. The planet Venus has two transits, separated by about eight years in a pattern which repeats every 243 years. The next pair will be in 2117, then 2125.[3] Mercury transits much more often, with a dozen or so per century (the next one due in 2032). There is only highly debated evidence of their observation by pre-telescopic astronomers, however by the 17th century astronomers had realized how useful the observation and timing of such events would be for refining models of the Solar System (see pages 210–211). The first recorded observations of a transit of Venus happened in 1639, and by the time of the 1761 and 1769 pair, it had been realized that geographically diverse observations of the exact time of the transit could be used (along with trigonometry, and the by then really excellent mathematical models of the orbits of Venus and Earth) to measure the distance between Earth and Sun. You can think of it as setting the base of a really long and skinny triangle, which once you measure the angle of, (which comes from the timing of the transit) you can obtain the height (distance to the Sun). This experiment was somewhat successful in the 18th century and refined by further adventures in 'transit expeditions' in the 19th century; today we have other methods to measure the distance to the Sun, but it remains an important milestone in astronomy.

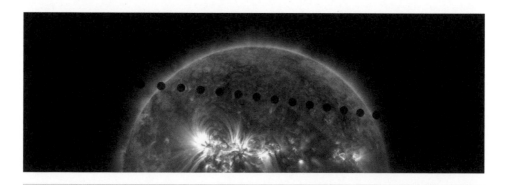

RIGHT

Venus Transit

Composite of images of the Venus transit taken by NASA's Solar Dynamics Observatory on 5 June 2012. The image, taken in ultraviolet light, shows a time-lapse of Venus's path across the sun.

---

[3] I was lucky enough to view the 2004 transit of Venus as a PhD student at Cornell, and tried to observe the 2012 transit from Portsmouth, UK, only to be foiled by cloud (in the best tradition of transit observers). These are the only two which will occur during my lifetime.

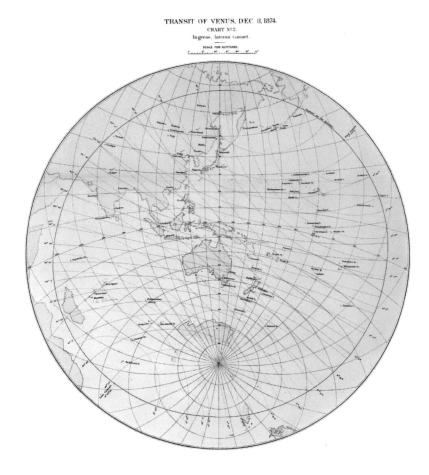

TRANSIT OF VENUS, DEC 8, 1874.
CHART N°2.
Ingress. Interior Contact.

SCALE FOR ALTITUDES.

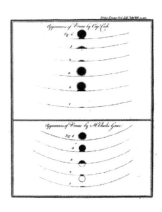

ABOVE LEFT AND RIGHT
Transit of Venus

The 1874 Transit of Venus chart, ingress 2 (interior contact). This transit took place on 8–9 December of that year.

Classic sketch of the passage of Venus as observed by Captain Cook, published in the *Philosophical Transactions of the Royal Society of London* in 1809 Illustrating what became known as the 'black drop effect' – an optical illusion which makes timing the exact start and end of the transit tricky.

Transit of Venus

With text in seven different Indian languages (Kannada, Malayalam, Marathi, Sanskrit, Tamil, Telugu and Urdu) as well as English, this short book was published in 1874, the year of the first of two transits of Venus in the 19th century (9 December 1874 and 6 December 1882). The author, Chintamanni Ragoonatha Chary (1822/1828–1880), was the first Indian member of the British Royal Astronomical Society. He was born in Madras in either 1822 or 1828 (records differ) and it is said that his ancestors had been great Hindu astronomers. He joined the Madras Observatory in 1840, initially as a low-wage labourer, being promoted to the role of 'astronomer' by 1864.

In his book, Chary urged Indian astronomers to be more interested in observing the transits themselves, and he provided details for a simple method to estimate the Earth–Sun distance from the observations.

4. Comets

Comets have long fascinated humanity. The word 'comet' comes from a Greek word meaning 'long-haired star', providing a descriptive name for how these look in the skies. Typically depicted as omens of bad fortune, their appearance in the skies is so different from everything else that when they are bright enough to be easily visible, they are a beautiful site. These 'Great Comets' (comets that are easily visible to the unaided eye) are fairly rare on human timescales, although common on astronomical ones.

Twelve such events happened during the 20th century. It can be hard to predict how bright an incoming comet might eventually be, which even today can lead to sensational press items, and misunderstandings (e.g. the rumour which spread that the 'green comet' of early 2023 was the first bright comet to appear in the sky for the last 10,000 years. This mixed up the appearance of any comet in the sky with the orbital period of this particular comet, which, while technically visible to the unaided sky from dark sites, was not particularly spectacular).

The modern astronomical understanding of a comet is a small lump of rock and ice, typically a few hundred metres to several kilometres across, which likely represent leftover material from the early formation of the planets in our Solar System. When such objects get close to the Sun, the ice is heated and begins outgassing, making the spectacular 'coma' and tail which are sometimes visible in our skies. The unpredictability of comets is caused by this process – it is not always clear how much gas will be released or how well it will reflect sunlight back to Earth.

Since bright comets might happen only a couple of times in any astronomer's lifetime, most published compilations of comets show drawings based on descriptions, or second-hand accounts from those fortunate enough to witness them first-hand.

RIGHT

The Bayeux Tapestry

Halley's Comet (as we know it today) depicted in the Bayeux Tapestry following its appearance in 1066.

Cometa orientalis

The year 1618 was unusual for comets, with three appearing in the skies visible to the unaided eye. The third, the 'Great Comet of 1618', was the most spectacular. Reports of its presence in the skies have been found from all over the world. One of these is *Cometa orientalis: kurtze beschreibung desz newen cometen*, which translates to 'The Eastern Comet: A Brief Description of the New Comet'. This short book was written by a German schoolteacher named Gothard Arthusius, who argued the comet was an omen of the coming of 'Judgement Day', including what he saw as clear evidence linking comets to various natural disasters. Such ideas have now been disproven; sadly disasters still happen regardless of the appearance of comets in the skies. However, the book contained a beautiful illustrated title page (see overleaf) so we will still save it for our library.

De cometis

De cometis (On Comets) is a book with some beautiful illustrations of comets, which was written by the infamous 17th-century English astrologer John Gadbury (1627–1704). It discusses comets mostly as omens, or portents of bad luck. The deceptively simple title is continued with the following ungainly description of the book's contents: 'a discourse of the natures and effects of comets, as they are philosophically, historically & astrologically considered. With a brief (yet full) account of the III late comets, or blazing stars, visible to all Europe. And what (in a natural way of judicature) they portend. Together with some observations on the nativity of the Grand Seignior.'

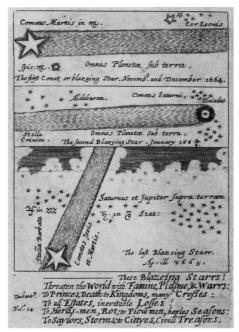

LEFT

De cometis

The title page and illustration of three comets from the book, published in 1665.

Image from the title page of
1618's *Cometa orientalis*.

The 'Great Comet of 1861' as
depicted by Edmund Weiss in
his *Atlas der sternenwelt* (Atlas of
Astronomy), published in 1892.

The first part of the book is about comets in general, while part two is a history of
great comets, and details of the events the author believes they were 'messengers of',
including a section on how comets (via astrology) impacted the lives of various notable
leaders (kings, emperors etc.) and the 'Grand Seignior' – the French term for an
emperor or sultan (of Turkey). This makes for a fascinating mixture of astronomy and
astrology, a common phenomenon at that time.

A Synopsis on the Astronomy of Comets

In this 1705 book, Edmond Halley (1656–1742), describes the detailed mathematical models for the orbits of 24 different comets that had been observed between 1337 and 1698, including his detailed observations of the comet that would come to be known as Halley's Comet, which he himself had observed in 1682.

He was able to apply the state-of-the art gravitational rules which had been published by Isaac Newton just a few years earlier (in 1687, see pages 200–205) to connect the comet he saw in 1682 with the historically recorded comet sightings of 1456, 1531, 1607 and 1682, and to predict a return of the comet in 1758.

The book itself is very mathematical in nature, with tables of figures and a diagram of geometry to help the reader understand how he predicted the comet's appearance in the skies. Here 'the astronomy of comets' is used to mean their motions in the skies; no comment is made on their 'astrophysics', or physical nature.

Probably most famous for his eponymous comet, Edmond Halley was an English mathematical astronomer and for the last 20 years of his life was the second Astronomer Royal in Britain (appointed in 1720 when he was 64 years old). His early career was quite adventurous. As a schoolboy he had written to the first Astronomer Royal to correct errors in the position of Jupiter, Saturn and some of Tycho Brahe's star positions. And he did not wait to finish his studies at Oxford: instead he arranged to travel to the southern hemisphere to map the stars. His work from this trip was used in two notable star atlases, *Firmamentum sobiescianum sive uranographia* from 1690 (see pages 41–44) *Atlas coelestis*, published in 1725 (see page 45–49). His behaviour didn't enamour him to the scientific establishment, but nevertheless he persisted with studies on gravity, comets, and a wide range of other astronomical topics. He famously predicted a return of 'his' comet in 1758, but he didn't live to see it; a fate many astronomers face when trying to view the comet, which only passes Earth every 75 years (the next being in 2061).

BELOW

A Synopsis on the...

Pages from *A Synopsis of the Astronomy of Comets* describing some of the mathematical tools needed to predict a comets appearance in the skies.

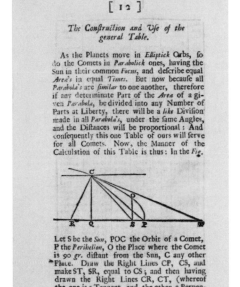

3

ASTRONOMY
AND CULTURE

In today's brightly lit and mostly indoor world, it is easy to forget that we live on a planet in space. But throughout most of human culture in the past, the connection between humans and the night skies was once much stronger. Our ancestors, in all parts of the world, were much more intimate with the night skies than most people today, so it should be no surprise that an impressive amount of our cultural practices have been impacted by astronomy.

Astronomy intersects and influences our cultures on a daily basis. Even that daily pattern itself has been set by astronomy – one day (or 24 hours) being the period of the revolution (or spin) of our planet Earth. Most of the time markers of the routines of humanity are set by our planet's various motions through space. As the Earth spins, it also moves in orbit around the Sun – a period that is encoded into our calendars as a year.[1] The length of a month, or 'moonth', is roughly the same as the period of a full cycle of phases of the Moon (29½ days, or a synodic month, to give it its proper name). The exact link has been lost to a variety of calendar tweaks or reforms which imposed months of 28, 29, 30 or 31 days to fit them exactly into a solar year, though the connection via the name remains. And while having seven days in a week is completely arbitrary in terms of the motion of the Earth, the names we use for the days are astronomical: there are seven days in your week only because of an astronomical fluke, where from here on Earth we cab see seven 'luminaries' in the sky – unusual, moving bright objects comprising the five planets, the Sun and the Moon. In Germanic and Old English, the names for Mars/Mercury/Jupiter/Venus sound like Tue/Wed/Thur/Fri respectively, while Saturday, Sunday and Monday more obviously come from Saturn, the Sun and the Moon.

There's nothing more important to human culture than the ability to have enough food to eat, and for much of human history, success or failure in being able to feed ourselves was

set by our ability to successfully navigate and plan agricultural work around the seasons. The four seasons, which are experienced by all cultures based in non-equatorial regions of the planet, are caused by the Earth having a spin (daily revolution) axis which is tilted a bit relative to the plane of our orbit around the Sun. That means that at some times of the year (we call them summer) the half of the Earth we live on tilts towards the Sun, and at others (winter), it tilts away. This impacts both the hours of daylight and the average temperatures, and early astronomers soon noticed how to connect the progress of these seasons to the patterns of stars they could see in the sky. Early humans had to be astronomers so they could decide when best to plant or harvest crops.

[1] One orbit takes 365¼ days, so in the modern world most years have 365 days, with every fourth (with some tweaks) being a leap year of 366 days, which works to keep the calendar aligned with the seasons.

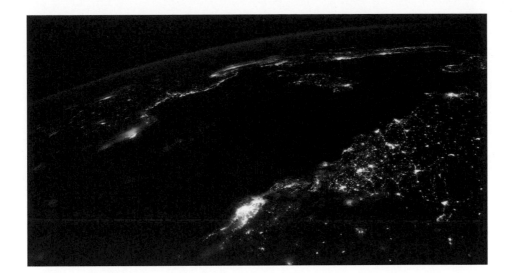

Perhaps because of the importance of the four seasons, many early astronomical charts are divided into four. These seasons are still marked by astronomical lodestones in our modern lives. Spring and autumn (or fall) are marked by the two 'equinoxes', literally 'equal night', days which occur around 21 March and September, when the 24-hour spin period of the Earth is broken equally into 12 hours of daylight and 12 hours of night everywhere on Earth. Many cultures still celebrate midsummer's night (the shortest night of the year), or 'the first day of summer'. This occurs around 21 June in the North – at a moment known as the summer solstice, literally 'the moment the Sun stands still, when the Earth sits at the point in its orbit which tilts the northern hemisphere most towards the Sun, and so the Sun momentarily appears to pause on its northward path through the constellations before turning to move back south. The final quarter is marked by the winter solstice, which at around 21 December is still close in date to some of the traditional northern winter festivals.[2] Even the dates between these quarterly orbit markers remain embedded into our culture. While their names as the 'cross-quarter days' are mostly forgotten, many people still mark at least some of them: in the USA, Punxsutawney Phil still gets to look for his shadow on 'Groundhog day', marking the midpoint between the winter solstice and spring equinox; 'May day', still celebrated at Oxford and other places by rising to meet the dawn, comes near to the midpoint between the spring equinox and summer solstice; and most of us have gone 'trick-or-treating' at Halloween, a date which sits near the midpoint between the autumnal equinox and the winter solstice. All of these festivals have their origins in the astronomical year.

Many of the great variety of religions here on Earth have also been inspired by astronomy. Some religions have made astronomical objects themselves into deities. Sun worship pops up in cultures across Africa, Asia, Europa and the Americas.

---

[2] Summer and winter are flipped in the southern hemisphere, with 21 June for the winter solstice and 21 December for the summer solstice).

And we should not forget that the major planets, in many languages, are named after various Greek/Roman gods. Even if not worshipping the skies directly, most religions use astronomical calendars. For example, lunisolar calendars – those that use a combination of the lunar and solar cycles to mark important dates – are remarkably common. Perhaps the most widespread of lunisolar-based celebratory dates is the Lunar New Year (in the West often called Chinese New Year, the Chinese refer to it as the Spring Festival), celebrated across most of East Asia. The Chinese lunisolar year dates back thousands of years, and the New Year usually sits at the second new moon following the winter solstice (once in a while the third new moon, for reasons similar to those which give the solar calendar leap days). The Jewish calendar also uses a lunisolar method, with dates of the Jewish high holidays for the most part (there are some rules about days of the week which are allowed) determined by the date of the new moon closest to the autumnal equinox (Rosh Hashanah) and Spring Equinox (Passover). Similarly, traditional Christian calendars use 'the Sunday after the first moon after the spring equinox' to set the date for Easter, and the date of Christmas is suspiciously close to the winter solstice. The Islamic calendar is perhaps the most directly tied to the Moon, still using observations of the first

OPPOSITE AND LEFT

Then and now

Today, astronomy seeps into our culture via iconic images from state-of-the art telescopes, like this colourful view of the 'Pillars of Creation', a region of active star formation in our Galaxy taken by the James Webb Space Telescope. NASA images are open source, meaning this image has been reprinted on all kind of objects, from earrings and clothing to phone cases and cushions.

From Sacrobosco's book *Compendium in sphaeram* (See Chapter 2), an illustration of how climate on Earth relates to the path of the Sun across the skies. The constellations of the zodiac are marked across the equatorial region showing how the Sun can pass directly overhead in these regions at different times of the year.

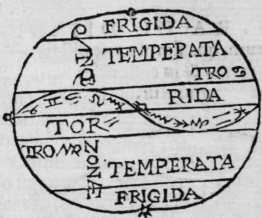

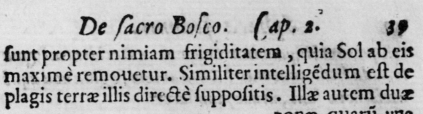

De ſacro Boſco. Cap. 2. 39

ſunt propter nimiam frigiditatem, quia Sol ab eis maximè remouetur. Similiter intelligédum eſt de plagis terræ illis directè ſuppoſitis. Illæ autem duæ zonæ quarũ vna eſt inter tropicũ æſtiualem & circulum arcticum, & reliqua, q̃ eſt inter tropicũ hyemalem & circulũ antarcticũ, habitabiles ſunt,& téperatæ, caliditate torridæ zonæ exiſtentis inter tropicos, & frigiditate zonarum, quæ ſunt circa polos mundi extremarum. Idem intellige de plagis terræ illis directè ſuppoſitis.

RIGHT AND OPPOSITE

Culture and observation

Lunar New Year celebrations
in New York in February 2023.
The date of the Lunar (or
Chinese) New Year is set by the
orbit of the Moon.

The Milky Way captured using
a long exposure photography
technique at night from Mount
Olympus, near Petrostrouga
in Greece. Due to artificial
lighting, only a small fraction of
people today live in places with
skies dark enough to see this
beautiful sight.

sliver of a crescent of the new moon to start each new month, so Islamic months therefore gradually cycle around through the solar calendar (the Islamic year being 10–11 days shorter than the solar year), meaning the holy month of Ramadan gradually moves backwards through the seasons year by year.

A belief that the motions of all objects in the night skies might directly impact their lives and so they might be able to predict the future by observing the motions of stars (aka astrology) has also popped up in many places. To be honest, there is clear evidence that at least some objects in the skies do impact life here on Earth. The motion of the Sun creates the seasons, the Moon creates tides, and plants are observed to follow the path of the Sun across the sky. For hundreds – if not thousands – of years astrology (literally 'interpreting the stars'), a practice which dates back to Ancient Mesopotamia, was not understood to be distinct in any way from astronomy (literally 'measuring the stars').[1] While modern professional astronomers no longer cast horoscopes, many of the historically famous astronomers you will encounter in this volume (Galileo, Kepler, even Newton) certainly did. Due to certain popular phone apps, astrology seems as popular today as it was when every monarch employed an astrologer, and throughout history much progress in our astronomical knowledge was funded by interest and belief in astrology. Most modern astrology is nothing more than a bit of fun, and science has moved on to demonstrate no connection between human affairs and the motions of the planets; however, in the modern world, most people still know something about their astrological 'star sign'. The twelve 'zodiac signs' (Aries, Taurus, Gemini, Cancer, Leo, Virgo, Libra, Scorpio, Sagittarius, Capricorn, Aquarius and Pisces) make an almost complete circle around the sky and are sometimes described as the constellation which

the Sun was in front of at the time of your birth. Astronomically, this connection to the apparent motion of the Sun in the skies has been lost due to the precession of the Earth's axis. This effect, which is exactly the same as that which makes a spinning top wobble, causes Earth's spin axis to gradually rotate around, shifting the position of the Sun relative to the background constellations in a cycle of roughly 26,000 years. So, for example, if you were born in January, you are traditionally a Capricorn, however, in the 21st century the Sun will be found in front of Sagittarius for most of January. To complicate matters further, modern astronomers added a thirteenth constellation to the path of the Sun – Ophiuchus, and keep discovering new minor planets.

In this section, we'll look at some of the books which record how astronomy impacted our cultures from all over the world. Many astronomers now fear a loss of this cultural heritage due to the skies becoming increasingly inaccessible. Humans have a tendency to develop a fear of the dark, and the ready availability of cheap electrical lighting – especially cheap, energy efficient, blue LED bulbs – is creating an increasing problem with light pollution. Even away from direct lights, in any location near to a city, the night skies are brightened by sky glow, obscuring the fainter stars. The vast majority of people now rarely see a truly dark sky at night. And even in dark sites, increasing numbers of artificial satellites are obscuring the natural skies. At the time of writing there are over 7,000 active artificial satellites in space, with more and more launched each year. These ancient books therefore preserve a precious set of moments in humanity's fascination with the skies.

---

[1] To be honest, based on my personal observations, and the number of times I have been introduced as an 'astrologer', astronomy and astrology are still frequently mixed up today!

1. Middle Eastern/Islamic Astronomy

The medieval Islamic world refers to a period of several centuries during the Middle Ages, sometimes called the Islamic golden age (roughly 8th–14th century). During this period the sciences, culture and the economy flourished in the Islamic world. Education was highly valued. Geographically, this Islamic Empire spanned from what is now the western edges of India to across the Middle East and Northern Africa right to the Atlantic, and it spread north to encompass most of what we now call Spain and south down the Arabic peninsula. In Chapter 1: Star Atlases you can see Al Sufi's *The Book of the Images of the Fixed Stars* recording Ancient Greek and Arabic constellations in 964. We'll collect some more examples of books from this period here.

The Important Stars Among the Multitude of the Heavens

One important record of knowledge in the medieval Islamic world comes in the form of the Timbuktu Manuscripts. Timbuktu, in Mali, near the western extreme of the Islamic world in north-west Africa, was an important stop for trade routes across the Sahara. Thousands of manuscripts were collected in this location, either created from local African knowledge or brought in from across the Arabic world. These manuscripts were kept safe for centuries, mostly in private households, but in recent years dramatic efforts have been made to both digitize them and smuggle them out of Timbuktu for safekeeping against modern jihadi forces.

RIGHT AND OPPOSITE

The Important Stars…

Written by Nasir al-Din Abu al-Abbas Ahmad ibn al-Hajj al-Amin al-Tawathi al-Ghalawi and first published c.1000 CE in Timbuktu, Mali.

The *Kashf al-Ghummah fi Nafa al-Ummah*, which is translated as *The Important Stars Among the Multitude of the Heavens* (sometimes known as *The Structure of the Heavens*) is one of these manuscripts, and appears to have been written around 1000 CE to train astronomers to understand the movement of the stars, presumably to calculate the dates to expect the changes of the seasons and to cast horoscopes. The book contains a single diagram, a circle divided in four which likely illustrates the four seasons or cardinal directions (North, South, East, West), and gives instructions for Ja'ar (presumably the student) to line up facing 'Dukur' (referring to the direction towards Mecca for prayers) in order to orient the diagram to the North.

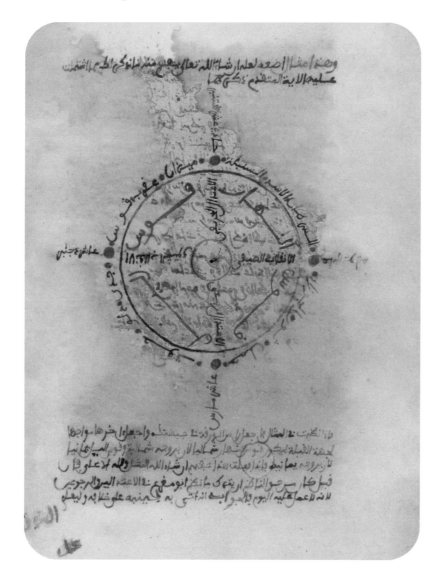

Written by Zakariya Ibn
Muhammad al-Qazwini
(1203–1283), the book was first
published in the 14th century
in Iran (in Persian as *Kitāb
'Ajā'ib al-makhlūqāt wa-gharā'ib
al-mawjūdāt*). The pages show
(left to right), phases of the
Moon, the geocentric model of
the universe and examples of
constellation artwork.

The Wonders of Creation

Kitāb 'Ajā'ib al-makhlūqāt wa-gharā'ib al-mawjūdāt (literally 'The Book of the Wonders of
Creatures and the Marvels/Oddities of Creation', but usually referred to as *The Wonders of
Creation*) was intended as a record of all 'the wonders of creation', including information
about astronomy as understood in 13th-century Iran.

The author, Zakariya Ibn Muhammad al-Qazwini (or al-Qazwini), lived in Qazvin,
a city in Iran. He was well educated and worked as a judge for many years before he
compiled his readings from a range of disciplines into his two books: *The Wonders of
Creation* and another book documenting geography and historical traditions. *The
Wonders of Creation* is particularly notable as an attempt to collect and summarize all
knowledge of both the heavens (astronomy) and more earthly considerations (animals,
planets, weather).

As was common at the time, he mixed together astronomical and religious (or
astrological) information. The first section contains the creation story, a discussion of
angels, and comparisons of calendars in the Islamic and Roman tradition to the Iranian
calendar. His model for the universe resembles that of Ptolemy – centred on Earth, with
concentric spheres for the planets, Moon, Sun and the fixed stars. Beautiful illustrations
of constellations are also included.

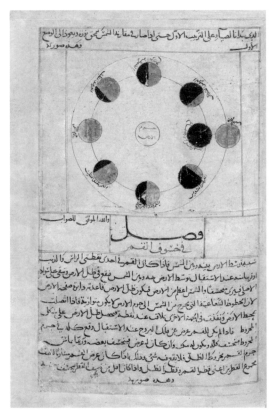

Al-Qabīsī's Treatise on the Principles of Judicial Astronomy

Al-Qabisi was a famous Islamic astrologer, astronomer and mathematician who lived in Syria (Aleppo) in the 10th century during the Islamic golden age.

Al-Qabisi's treatise on 'Judicial Astronomy' is really about astrology, but at the time there was no separation between astronomy and astrology. The book was so popular that it was copied frequently, and even today there exist tens of different Arabic copies, and hundreds of translations into Latin. Judicial astrology is a now obsolete term which was used in the Middle Ages to refer to the parts of astrology that concern themselves with predicting human behaviour. At this time, astrology was split into 'natural astrology' (e.g. predicting the weather, as well as medical ailments) which was considered reasonably scientific, and 'judicial astrology' (i.e., everything else) which was later considered heretical by the Catholic Church.

This edition of the book, published in Lyon, France around 1519, is translated into Latin and contains a commentary of the work by John Danko of Saxony, a medieval European astronomer who lived in the 13th–14th century, mostly in Paris. John of Saxony spent much of his career translating and explaining the work of ancient astronomers, and he is notable for his work based on this book by Al-Qabisi as well as translations from other Islamic astronomers and Ptolemy's *The Almagest* (see Chapter 4: Developing Our Model of the Universe).

RIGHT AND OPPOSITE

Judicial Astronomy

Credited to John Danko of Saxony (1327–1355), this edition was published in France around 1519.

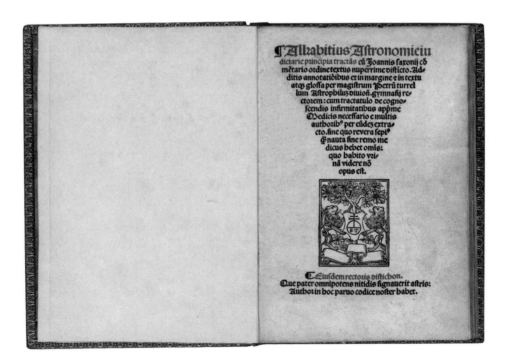

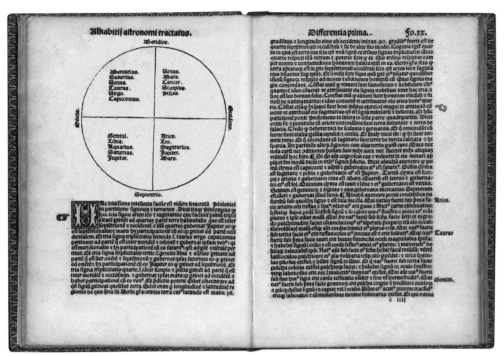

2. Indian Astronomy

Astronomy in India flourished from perhaps as early as 1500 BCE, through to what many would call its culmination in the Kerala School of Astronomy and Mathematics, active in the 14th–16th century. The geographic location of India enabled links with other astronomers from Greece, China and the Islamic Empire, but Indian astronomy also developed independently, with notable contributions to spherical trigonometry and a unique Hindu calendar based on a geocentric model of the universe with the start of the year linked to the winter solstice.

Vedanga jyotisha

An early text on Indian astrology, quite possibly the oldest, although the exact date of publication is debated, the *Vedanga jyotisha* is the foundational text for the Vedanga (Hindu) discipline, known as Jyotisha, which focuses on astrology and timekeeping (sometimes known as Hindu astrology, or Indian astrology). Scholars believe the first publication was some time in 1200–1400 BCE.

BELOW AND BELOW RIGHT
Hindu astronomy

Some of the stunning remains of the 18th-century Jantar Mantar Observatory in Jaipur.

A colourful example, showing the Hindu Calendar, from the Library of Congress in the USA.

Two copies of the book remain today, each slightly different in detail, but both contain information about how to use the movement of astronomical objects (the Sun, Moon and planets) for timekeeping, along with mathematical details of how to use trigonometry and other methods to predict the motion of the planets from their orbits. It is often mentioned that this book contains a count of up to 4.32 billion years for the longest of the astronomical cycles mentioned. This is notable as such a long period, and much closer to current understanding of the age of the universe (about 14 billion years) than any other contemporary astronomers had reached.

Aryabhatiya

A book of astronomy and mathematics presenting the work of the Indian mathematician Aryabhata (476–550 CE) was published some time in the 5th century in India. The work describes a geocentric model for the Solar System, with details to calculate the motion of the planets. Aryabhata notably described that the reason for the daily motion of the stars was due to the rotation of Earth. He also reasoned that the Moon and the planets must shine from reflected sunlight, explained that lunar and solar eclipses are due to shadows either from or falling on Earth respectively, and calculated the lengths of the sidereal day and year very precisely.

Tantrasamgraha

The *Tantrasamgraha* (literally 'A Compilation of the System') is a description of the state-of-the art of Hindu Astronomy from Kerala School of Astronomy and Mathematics in the early part of the 16th century. Several copies of the manuscript survived to the modern era, mostly written on palm leaf in Malayalam script.

The author, Nilakantha Somayaji (1444–1545), was an Indian astronomer and mathematician. He is most famous for his work on this book, which was first published around 1501.

Alongside the description of the astronomy of the time, this book is notable for its mathematical sophistication, describing series expansions for trigonometric functions which would be familiar to all advanced mathematics students today. This kind of advanced mathematics is essential for advances in astronomy (see Chapter 4: Developing Our Model of the Universe), which by its nature, has to deal with the complex geometries of spherical trigonometry (angles and distances on spheres). Somayaji used these and other techniques to refine the model for the motion of the planets Mercury and Venus. The model used was hybrid geocentric model, in which all the known planets (excluding Earth) orbit the Sun, while the Sun orbits Earth.

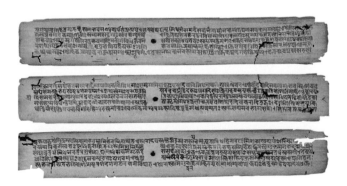

3. Mesoamerican Astronomy

Mesoamerica refers to a historically significant geographical and cultural area which crosses the southern part of North America, and much of Central America. The pre-Columbian peoples in this region had flourishing civilizations, some of which lasted for thousands of years, developing sophisticated scientific, religious and cultural practices, and building large cities. Spanish colonization of this region began in the late 15th century, and was particularly active in the 16th century. This European invasion is estimated to have killed something like 80 per cent of the native population, mostly via the spread of infectious diseases, although forced labour, slavery, resettlement and military violence certainly contributed. In our Astronomers' Library we'll collect two books which record astronomical knowledge from Mesoamerican peoples, one from the Maya, and one from the Nahua (more commonly, but incorrectly, known as the Aztecs).

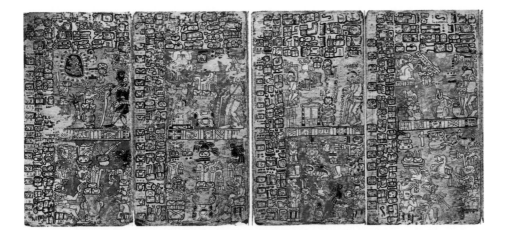

The Madrid Codex

Many texts written in this part of the world were taken to Europe during and after colonial activity. Also known as the *Tro-Cortesianus Codex* or the *Troano Codex*, *The Madrid Codex* is a pre-Columbian Maya book.

The Dresden Codex

In the astronomical world the Maya are famous. The Maya are a people who live in an area which is now split between south-eastern Mexico, Belize, Guatamala and western parts of Honduras and El Salvador. Maya traditions of astronomy are well known for their sophistication, particularly in their development of the calendar, records of the motions of planets, and predictions of solar and lunar eclipses.

Codices (or in the singular, a codex) are the traditional type of book from the pre-Columbian Maya civilization. They are printed on traditional Maya bark paper (known as huun-paper) and folded zig-zag style into books of many pages. When the Spanish invaded Maya in the 16th century, an uncounted number of these books were burnt, a violation of Maya culture which resulted in an almost indescribable loss of human knowledge. Only four codices survived this period, and these have been named after the locations they were rediscovered, namely Dresden, Madrid, Paris, and Grolier thus also erasing the connection to Maya culture in the titles. All four of these remaining codices include astronomical information, although *The Dresden Codex* is generally considered to be the most important, and until recently the oldest surviving book from the Americas. It recently emerged that *The Grolier Codex*, renamed *The Maya Codex of Mexico*, may be older. Nevertheless, it is *The Dresden Codex* we will collect in our Astronomers' Library.

The exact date of the origin of the codex is unknown. Opinions on the dating differ among experts, but it is believed to have been written somewhere between the 12th and 14th centuries in or near Chichen Itza, a Maya city in the Yucatan Peninsula with a famous observatory. It is speculated that the codex was sent to Europe in 1519 by the Spanish invaders as a gift of items representing Maya culture to the Holy Roman emperor Charles V. Why it was not destroyed alongside so many other Maya codices is not recorded. By 1739, the codex had made its way into private ownership, when a librarian from Dresden purchased it. It has remained in Dresden ever since, sadly suffering significant damage during the Second World War bombing of the city.

The codex contains a number of astronomical tables detailing the motions of Venus and the Moon, and it predicts the upcoming dates of eclipses. Maya numbers were recorded with sequences of bars and dots in base 5 (i.e. one bar is 5, two are 10), and much of *The Dresden Codex* is made up of tables of these numbers interspersed with beautiful illustrations and Maya hieroglyphic writing.

The Maya Long Calendar is actually somewhat notorious to modern astronomers, partly because of how it was indirectly credited for the (erroneous it turned out) predictions of the end of the world on 21 December 2012. This date marked the end of the 12th *b'ak'tun* (long count) in the Maya calendar. In the lead up to this date, the internet exploded with misinformation about it being the end of the world, especially because (as the story went) it coincided with the winter solstice as well as an alignment of the Sun with the Galactic centre. There is absolutely no evidence the Maya believed the end of their 12th long count was the end of the world, and it was likely merely a coincidence that it lined up with the solstice. The precession of Earth makes the Sun's position cross the Galactic equator periodically – in 2012 it was close to halfway across it,

Part of the solar eclipse tables showing a serpent eating the symbol for 'day'.

Excerpt showing a table counting the days between Venus appearing as a morning star, disappearing behind the Sun, being an evening star, then going behind the Sun again. The dots and bars of Maya numerals are slightly obscured in this copy, but they add up to 584 days, very close to the modern timing of the cycle (583.92 days), which measures the number of days it takes for Venus to overtake the Earth on their orbits around the Sun (i.e. the separation in times between Venus passing between Earth and the Sun).

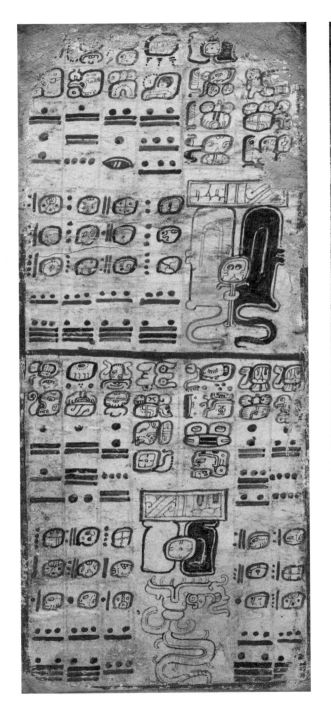

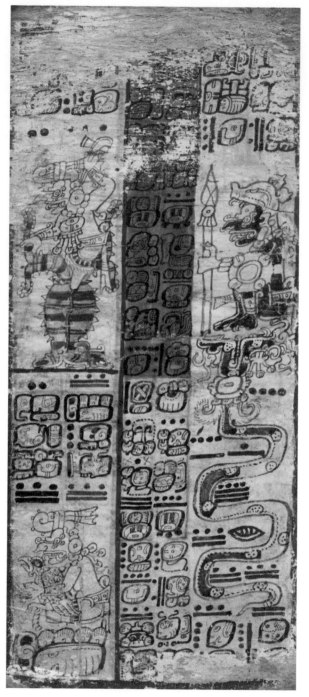

although it takes several hundred years to cross completely. In any case, I am writing this more than a decade after the world definitely didn't end in 2012.

The Maya religion included many gods, but Itzamna, the sun-god, was one of the most important. This meant the Maya spent a lot of effort tracking the motion of the Sun, and *The Dresden Codex* includes tables of dates of solstices and equinoxes which demonstrate that Maya astronomers knew that a complete year was about a quarter of a day longer than their traditional 365-day *Haab'* (one of the Maya calendar years). As well as their astrological use, tables tracking progress of the seasons would have been useful for agriculture.

To the Maya, Solar eclipses (when the Moon passes between the Sun and Earth) were crucial events to predict, as (in a belief which echoes those in many other cultures) they were thought to be dangerous events which might bring catastrophes in their wake. This is represented in *The Dresden Codex* with a serpent eating the hieroglyph for 'day' (see page 123). The tracking of lunar eclipses (when Earth passes between the Sun and the Moon) is also highly accurate in *The Dresden Codex*.

In Maya astronomy, Venus (*Noh Ek'*, or 'the Great Star') was considered the second most important object in the sky. Unlike many early astronomers who often considered the 'evening star' and 'morning star' appearances of Venus as two separate objects, the Maya realized it was a single planet and recorded that it has a pattern of 584 days for a full cycle of its appearances in the evening, morning and disappearances into the glare of the Sun (every 584 days Venus passes between Earth and the Sun).

The Dresden Codex also contains similar tables tracking the motions of the planets Mercury and Mars, including noting the timing of Mars's retrograde motion (when it appears to move backwards relative to the fixed stars for a time). This is a significant astronomy book to include in our Astronomers' Library, and we can only imagine what other records of Maya astronomy were lost in the post-Columbian period.

OPPOSITE AND BELOW
The Dresden Codex

A wealth of fascinating and practical information is stored in the codex, indicating a highly advanced civilization.

Historia general de las cosas de la Nueva España

The Nahua are an indigenous people from Mesoamerica, from regions of what are now called Mexico, El Salvador, Guatemala, Honduras and Nicaragua. The 'Aztecs' and Mexica were both of Nahua ancestry. The *Historia general de las cosas de la Nueva España* (General History of the Things in New Spain) is an enormous encyclopaedia-style book which provides an extensive record of Nahua culture in the 16th century.

The *Historia* was the culmination of what today might be called a research project, or ethnography, led by Friar Bernardino de Sahagún (1499–1590), a Franciscan friar and missionary famous for his travels around Mesoamerica during the 16th century. He first travelled to South America when in his early thirties and spent the rest of his life there – over 60 years. He became fluent in Nahuatl, the language of the Nahua people (including Aztecs), and documented their culture.

Sahagún originally travelled to 'New Spain' (now Mexico) as a Catholic missionary, funded by the Spanish invaders of Mesoamerica. Motivated by a desire to convert the locals to Catholicism, he worked with Nahua students in what is now called Mexico City, at the first European-founded higher education site in the Americas. He later travelled widely across the rest of Mexico, working with some of his former Nahua students to document local cultures.

The *Historia* covers a vast array of cultural information about the Nahua peoples at that time, from religion and ritual practices to economics and information about agriculture. The pages of the book generally contain text in both Nahuatl and Spanish.

The Nahuatl text, written by Nahua men whose names are not recorded, was a translation into the Latin alphabet of what was previously a hieroglyphic style of writing. Sahagún reviewed these entries and wrote the Spanish translation. The book also includes over 2,000 beautiful illustrations by native artists whose names are also not recorded. The most complete existing version of this manuscript was rediscovered in Florence, Italy in the 18th century so the book is often referred to as *The Florentine Codex* – a name given to it by 19th-century American (English) translators, in the tradition of other scholars of Mesoamerican ancestry at the time. This serves to erase its Mesoamerican origins.

The reason we include the *Historia* in our Astronomers' Library is because it includes significant information about astronomical knowledge and practices of the 16th-century Nahua. Two of its 12 volumes contain some astronomical information. *Book 4: The Soothsayers* is mostly about omens and fortune-telling, and as such includes some information on astrology. *Book 7: The Sun, Moon and Stars and the Binding of the Years* is entirely about Nahua astronomy, including information about their creation stories and calendars.

Book 7 includes the Nahua legend of the fifth creation of the Sun and Moon. The Nahua believed the world had previously gone through four cycles of creation and destruction, and that we live in the fifth one. The story is used to explain why the Nahua believed the outline of a rabbit could be seen on the face of the full moon. We know today that the patterns on the full moon are caused by alternating bright lunar highlands, and dark spots from the flatter areas, but cultural mythology giving reasons for these patterns are quite common across the world. The Nahua story starts with the gods wanting to create the next Sun. The gods got together and decided that from among them, Tecciztecatl, who was wealthy and strong, should become the Sun, while Nanahuatzin, who was poor and ill, was rejected. However, for this to happen Tecciztecatl was required to sacrifice himself by leaping into a fire. When the time came Tecciztecatl was too afraid, but Nanahuatzin leapt in willingly. Tecciztecatl then followed him in shame. This caused two suns to appear in the sky for a time, but the rest of the gods were either so disgusted by Tecciztecatl's cowardice, or scared of their being two suns in the sky, that they threw a rabbit in the face of the second sun, and it instead became the Moon.

ABOVE LEFT AND RIGHT

Historia general…

Two suns from Book 7, perhaps from the culture's creation story, which involves two suns in the sky for a time.

Like the Maya, much of Nahua astronomy was focused on predicting eclipses – when the Moon covers the Sun. This page includes an illustration of what appears to be a solar eclipse.

Historia general…

Book 7 includes beautiful illustrations of a comet, clouds and a rainbow, alongside a description of weather patterns. At the time it is likely that astronomers believed comets were atmospheric phenomena like clouds and rainbows.

Book 7 also includes explanations of how the Nahua counted the year, and the sign for each year. It ends with a description of the 'Binding of the Years', sometimes also called the 'New Fire Ceremony'. This was a ritual performed every 52 years when the Nahua long calendar ended and was said to be necessary to prevent the end of the world. It was rather gruesome. All fires were put out, and then to begin again a new fire was started in the body cavity of a captive as illustrated here. The timing of this event was tied to astronomy and the appearance of particular constellations in the skies.

Libro. 7.º

de la Astrologia

y philosophia

la tomauan, por buen aguero: y al principio del tiempo. que començaua a aparecer por el oriente.

¶ Capitulo quarto de las cometas.

Llamaua esta gente, a la cometa, citlalin popoca. que quiere dezir, estrella que humea: teni an la por prenostico de la muerte. de algun principe, o rey, o de guerra, o de hambre. La gente vulgar. desia esta es nuestra hambre.

A la inflamacion de la cometa, llamaua esta gente, citlalin tlamina, que quiere dezir, la estrella tira saeta: y desi an, que siempre, que aquella saeta, caya sobre algu na cosa biua, liebre, o conejo, o otro animal. y donde he ria. luego se criaua, vn gu sano. Por lo qual aquel ani mal, no era de comer: por esta causa, procuraua esta gente, de abrigarse de noche, por que la inflamacion de la cometa no cayese sobre ellos.

çotica in mamalti, contlatzitzi cuiniliaia, contlatlatlaxiliaia, contlaiiauiliaia.

¶ Inic naui Capitulo, itechpa tlatoa: in cicitlaltin.

¶ Citlalin popoca.
Mitoa: tlatocatetzauitl, ic tlato camicoaz, aço aca uey tlaçopilli iemiquiz: yoan no quitoaia, aço cana icoalmotzacoaz, aço ie oli niz teuatl tlachinolli: yoan ano ço iemaianaloz. Quitoaia in ma ceoalti: aço tapiztli, aço apiztli quitoa.

¶ Citlalin tlamina,
Mitoa: amo nenquiça, amo nē uetzi, in tlaminaliz: Haocuillo tia. Auh in tlamintli, mitoa citlalminqui, ocuillo, aocmo qua lo, mauhcaitto, tlaclitto, bibielo, tetlaieltia. Auh in iooaltica vel nemalbuilo; neolololo, netlapa cholo, nequentilo, netlalpililo: imacaxo in itlaminaliz citlalin.

El arco del cielo, [es] de arco de canteria parencia, de diuerso quando aparece de serenidad.

¶ Aiauhcocamalotl.
Iuhquin uitoliuhqui: tlauitol
tic, coltic, inic oalmoquetza,
tlatlatlapalpoalli, motlatlapal
poub initlachieliz. In centla
mantli tlapalli, itech neci: xo
xoctic, quiltic, quilpaltic, iia
paltic, quilpalli, iiapalli: yoan
coztic, xopaltic, xochipalli: ni
man ie chichiltic, tlapaltic: yoa
tlaztaleoaltic, tlaztaleoalli :
yoan texotic, texotli, matlal
tic, matlalli. Auh quitoa , in
icoac oalmoquetza: quinestia,
quiteittitia, quinezcaiotia, ic
macho, ic machizti, ic itto: in
amo quiauiz, amo tlaelquiauiz,
amo tilaoaz: can quimomoiaoaz
in mistli, quipopoloa, quehtel
tia, quiiacatzacuilia inquia
uitl , in tlaelquiiauitl. in tepal
tili, in techacoani, in tecoqui
tili. Intla cenca omotlatlali mis
tli, inouel cuiauicheoac, inono
uiian tlatlaiooac: can quipo
popoloa. Intlanel quiaui, aoc
mo cenca tilaoa, aocmo mol
huia : ca aoachquiaui, aoach—

de pino, luego, lleuauanla al tem
plo del ydolo de vitzilobuchtli, y
ponjanla , en vn candelero, hecho
de aluçanto, puesto delante del
ydolo: y ponjan enel. mucho en
cienso de copal. I de alli tomaua
y lleuauan al aposento, de lossa
cerdotes, que se dize mexicanos: y
despues cotos aposentos, de los di
chos mjnistros. de ydolos: y de alli
romauan, y lleuauan todos los
vezinos, de la ciudad. I era cosa
de ver, a aquella multitud
de gente, que venjan por la
lumbre: y ansi hazian hogue
ras grandes, y muchas, enca
da barrio, y hazian muy gran
des regozijos.

A. ¶ Lo mesmo hazian, los otros
sacerdotes, de otros pueblos:
porque lleuauan, la dicha lu
bre, muy apriessa, y aporfia

ic motlaloa: ipampa inic iciuh
ca, caxititiuetzizque tletl imal
tepeuh ipan: ca achtopa ic ne
nemachtiloia, muchichioaia
in tlecuioani: itoca tlepilli. Auh
ie choatl, ic quioalaxitiaia in
tlenamacaque : oc ie achto, vm
pa quitlecauiaia, quitlamela
oaltiaia iniicpac teucalli: in
vmpa mopieia ixiptla vitzi
lobuchtli, tlequazco contlalia
ia : nima ic contepeoa, conto
xaoa iniztac copalli. Auh ni
man ic oaltemo, oc ie no ach
to, vmpa quitqui, quitlameca
ualtia in calmecac, itocaioca
mexico: ic çatepan moiaoa, tle
tletlalilo, innouiian cacalmeca
cacalpulco: niman ie ic iauh, in
nouiian tetelpuchcalli. Ic vn
can inisquich onxoquiui, on
motepeoa, ontapaliui maccoal
li, in motlecuilia: icoac ic noui
ian, tepan moiaoatiuetzi intletl
ne tletletlalilo, neioiollalilo. Ca
no iuh quichioa, inisquich alte
peoa tlenamacac: inic quit
quia, quinenemitiaia tletl, ca

4. Europe in the Middle Ages

In Europe, the period following the fall of the Roman Empire until the 'dawn' of the Renaissance is known as the Middle Ages, or medieval period (approximately 500 CE to 1500 CE). While culture and knowledge prospered during this time across much of the world, in Europe, historically this has a reputation as a time of 'ignorance and superstition'. The collapse of central authorities alongside population decline are commonly cited as the reasons for the gap in progress and lack of scientific thought. However, some modern historians argue that levels of superstition and ignorance may have been overstated by the cultural and scientific leaders of the Renaissance as they sought to distance themselves from their predecessors. It is certainly the case that notable books remain from this time which record the astronomical (and astrological) knowledge of the period, and not all of it is about superstition – there are books that reveal a clear search for knowledge and understanding.

De responsione mundi et de astrorum ordinatione

De responsione mundi et de astrorum ordinatione (On the Explanation of the World and the Arrangement of the Stars), sometimes called *De naturae rerum* (On the Nature of Things) is a book documenting late Roman Empire/early medieval Spanish knowledge of astronomy. It was written by Isidore of Seville (c.560–636), a Spanish scholar and Catholic priest from the early Middle Ages. Isidore of Seville was born into a family of Spanish–Roman nobility right at the end of the Roman Empire. Many historians refer to him as 'the last scholar of the ancient world', and much of his most famous work was focused on collecting classical texts, which might otherwise have been lost in the Middle Ages.

De responsione mundi et de astrorum ordinatione was a popular source of learning about astronomy throughout medieval Europe, and it was adopted by some in the Catholic Church as an important measure of the age of Earth. Copies of it from the 15th century were printed in ink, with hand- decorated initials and illustrations (see overleaf), and are among some of the earliest European printed records of astronomy. This book was written at a time and location when there was little, if any, distinction between astronomy and astrology. In particular, medical science at the time was driven by astrological beliefs about the importance of understanding the different human temperaments (or humours) and their interaction with the four elements (fire, earth, water and air). Medical astrology was a significant field, taught in the first universities, which developed during this period. Prospective doctors would have to learn how to calculate the locations of the planets at the time of a person's birth and compare them to the pattern which happened/would happen at the moment of a significant event. The book includes diagrams explaining these beliefs and calculations, along with details

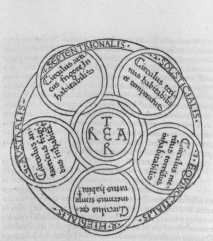

Ideo autem equinoctialis circulus inhabitabilis est· qa sol per medium celum currens nimium his locis facit feruorem· ita ut ne fruges ibi nascantur· propter exustam terram· nec homines propter ipm ardorem habitare permittantur. At contra septen/trionales & australes circuli non habitantur· quia a cursu solis longe positi nimio cch frigore ventorumqz gelidis flatibus con tabescunt. Solsticialis vero circulus qui in oriente inter septen/trionalem & estiualem est collocatus· vel iste qui in occidente inter estiuum et australem positus est·ideo temperati sunt·eo op ex vno circulo frigus ex altero calorem habeant. De quibus vir gilius. Has inter mediamqz due mortalibus egris. Concesse sunt munere diuum. Sed qui proximi sunt estiuo circulo ipsi sunt ethiopes nimio calore perusti. De Partibus mundi ¶ Ca·XII·

Artes mundi sunt quatuor·ignis·aer·aqua·terra· qua/rum haec est natura.Ignis tenuis acutus ac mobilis. Aer mobilis acutus & crassus. Aqua crassa obtusa & mobilis.Terra crassa obtusa immobilis. Quae &iam ita sibi inuicem cõmiscetur. Terra quidem crassa obtusa & immobilis cum aquae crassitudine & obtusitate conligatur.Deinde aqua aeri in crassitudine & mobilitate coniungitur. Rursus aer igni communione acute & mobili conligatur.Terra autem et ignis a se separantur sed a duobus mediis aque & aere iungunt. Hec itaqz ne confusa minus conligatur subiecta expressa sunt figura

Ceterum sanctus. Ambrosius hec elemẽta per qualitates qbus sibi inuicem quadam nature communione cõmiscentur·ita his verbis distinguit.Terra inquit arida & frigida est. Aqua frigida atqz humida est. Aer calidus & humidus. Ignis calidus est

of the division of time into hours, days, weeks, month and years. There is a diagram showing the arrangement of the geocentric Solar System with circles for 'Luna, Mercurius, Lucifer [the morning star – or Venus], Sol, Vesper [the evening star, which here presumably means Mars], Foethon (Jupiter) and Saturnus', along with charts of the phases of the Moon and a drawing of the Sun.

ABOVE AND LEFT

Di responsione mundi…

Pages from the book, written by Isidore of Seville and probably first published around 600 CE in Spain.

A painting of Isidore of Seville from around 1700. It hangs on the wall of the cathedral in Seville, Spain.

Di responsione mundi…

These pages are from an edition
of the book printed in Germany
around 1470.

Diagrams to do with
timekeeping and/or astrology,
including two charts showing
the 12 months of the year
around a circle (top left and
lower right). A chart labelled
with the four cardinal directions
(upper right), and the four
medieval elements (fire, air,
water and earth) as they were
thought to link to people's
temperaments through the year
(lower left).

A page showing the medieval
geocentric model of the
universe, with Earth at centre,
and concentric circles of
the Moon, Mercury, Venus
(Lucifer), the Sun, Mars
(Vesper), Jupiter (Fofton)
and Saturn.

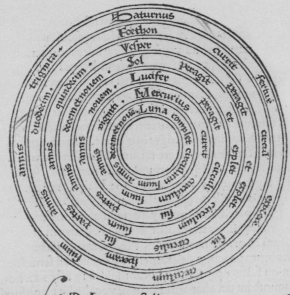

¶ De.Lumine ſtellarum · Cca.xviii

Tellas non habere ͵pprium lumen· ſed a ſole inlumi
nari dicuntur. Nec eas vnquã de celo abſcedere. Sed
veniente ſole celari. Omnia enim ſidera obſcurantur
ſole oriente ſed non cadunt. Nam dum ſol ortus ſui ſigna ͵pmi/
ſerit·omnis ſtellarũ ignis ſub eis luminis fulgore euaneſcit. Ita
ut ͵pter ſolis ignem nullius ſideris ſplendor videatur.Hinc etiã
& ſol appellatus eo ꝙ ſolus appareat obſcuratis cũctis ſiderib9
Nec mirum hoc de ſole intelligere cum luna plena & tota nocte
fulgente pleraꝗ aſtra non luceant.Eſſe autem ſtellas ͵p diem in
celo ͵pbantur ſolis deliquio· ꝙ qñ ſol obtecto orbe lune fuerit
obſcuratus· clariora aſtra in celo v. deantur. Stelle autem ſcdm·
miſticũ ſenſum ſancti viri intelliguntur. De quibus dictũ eſt·
Qui numerat multitudiné ſtellarũ. Sicut eni omnes ſtelle a ſo/
le illuminant.Ita ſancti a criſto gloria celeſtis regni g̶lorificanť
Et ſicut ͵p fulgore ſolis & vim maxima luminis eis ſidera obtũ/
duntur.Ita & o̅ns ſplendor ſanctorũ in cõparatione glorie xp̅i
quodam i̅o obſcuratur· & quemadmodũ ſtelle ſibi defferunt

De temporum ratione

De temporum ratione (or The Reckoning of Time) was written in 725 CE by the English monk, Bede (c.673–735). Sometimes known as Bede the Venerable, he was a Catholic monk from the Kingdom of Northumbria, a region which is now part of northern England. He was sent to the monastery at around the age of seven and spent most of his life there. Bede became a great teacher and scholar, an achievement which was recognized in 1899 when the then Pope named him a 'Doctor of the Church' – a type of Catholic Saint particularly noted for contributions through study, teaching or writing. Only 37 such saints are recognized, and Bede is the only native-born Briton among them.

Along with this book, Bede is also famous for his work on the *Historia ecclesiastica* (Ecclesiastical History of the English People). Perhaps as he worked on the *Historia*, an important record of English medieval history, he developed an interest and skill in timekeeping and calendars using astronomical methods. He had a particular focus on 'computus', or the calculation of the date of Easter, which, based as it is on a lunisolar method (the first Sunday after the first full moon after the spring equinox) requires calculations of both the phase of the Moon and an accurate solar year.

Even before his treatise on The Reckoning of Time, Bede's work was impacting calendars and timekeeping across Europe. He popularized the idea of counting years from the birth of Christ. At this time, different regions often used different methods of counting the year, and while Bede wasn't the first person to suggest the 'Anno Domini' method (now usually referred to as 'the Common Era', or CE), his work led to the wide adoption of it, something we are still using today.

In *De temporum ratione*, Bede brought all of his knowledge and understanding of timekeeping and calendars together into a single book. This was in large part an attempt to reconcile astronomical knowledge of timekeeping with details of the chronology of history in the Christian Bible. It includes a comparison of various calendars he was aware of (including the Jewish calendar and those of the Egyptians, Romans, Greeks and

RIGHT AND OPPOSITE
The Reckoning of Time

Pages from the book, written by Bede and first published in England, c.725 CE. Left: an infographic showing the movement of the Sun and planets in the sky over time. Centre: an illustration of how to count using finger symbols (sometimes called finger reckoning). The numbers are also indicated with Roman numerals and the arm symbols can be used for estimates of large numbers. Right: a beautifully decorated page.

'Saint John Dictating to the Venerable Bede', from an ancient manuscript, highlighting the regard in which Bede was viewed.

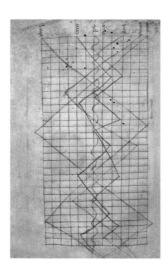

Anglo-Saxons). A large section of it is devoted to methods for the reckoning of the date of Easter, which included calculations of the date of the full moon, and the motion of the Sun and Moon through the zodiac constellations. The book ended with Bede's chronology of Earth from the time of Adam, and a discussion of prophecies about the second coming of Christ and Judgement Day. In this section he provided a new estimate of the age of Earth, calculating 3952 BCE as the date of creation. This got him into trouble in some circles, who believed it was heretical to criticize the date previously published by Isidore of Seville (see pages 130–131).

Alongside these details about timekeeping, calendars and Christian understanding of the history of Earth, the book also contains some information on physical astronomy. For example, Bede comments on the spherical shape of Earth, describing it as being 'not circular like a shield or spread out like a wheel, but [it] resembles a ball being equally round in all directions'. Bede used this, along with observations of the changing path of the Sun through the seasons, to explain the changing hours of daylight, which are certainly highly variable in the high latitudes of England, where he lived. Bede also gave an explanation for the timings of twice daily

high tides as being related to the daily motion of the Moon, as well as how the height of tides changed with the seasons related to the changing path of the Moon across the skies.

This was a popular book for centuries in Europe, perhaps because it was written as a textbook and included plenty of examples for teachers to discuss with their pupils. It also provides a priceless record of astronomical knowledge among the most educated medieval Anglo-Saxons.

5. East Asian Astronomy

Historical records from ancient Asia, mostly in the form of scrolls, reveal a rich, highly educated and diverse society. As the largest country in the region by far, ancient Chinese astronomy gets particular mention; other countries, like Japan and Korea were heavily under the influence of China for much of their history, but they still developed their own cultural practices around astronomy. We noted in Chapter 1: Star Atlases that the oldest paper star atlas comes from Tang dynasty (7th and 8th centuries) China. The oldest observatory in East Asia, the Cheomseongdae (see overleaf), is found in South Korea, dated to around the same time. And the Cheonsang Yeolcha Bunyajido, a stone-carved star map from Josean dynasty Korea (14th century) provides a record of traditional Korean star patterns (see page 20). Astronomical texts of note also remain from Edo period (17th century) Japan.

Divination by Astrological and Meteorological Phenomena

Early writing materials in China included the use of bamboo pens to write on wood. These were likely heavy and inconvenient to transport, and the development of hair brushes led to a period where it became more common to write directly onto silk materials, although this was expensive. Subsequently, the Chinese court official Cai Lun (c.60–121 CE) is credited with the invention of wood pulp-based paper in the 1st century CE. Nevertheless, in our Astronomers' Library it seems important to include one of the traditional Chinese silk manuscripts. *Divination by Astrological and Meteorological Phenomena*, often dubbed simply the Book of Silk, which was originally published during the Western Han dynasty (c.200–0 BCE), is our choice. This book is a pictographic record of 29 different comet sightings (or *hui xing* – literally 'broom stars') recorded by Chinese astronomers over a period of 300 years. The text accompanying each illustration refers to how each comet was believed to link to particular events on Earth, treating them as omens, or harbingers of doom such as plagues, warfare or meteorological disasters. Today it is still regarded as an astronomically significant text as the first clear atlas of comet sightings.

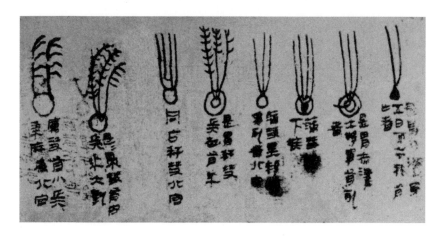

Dunhuang Star Chart

Of particular note in an astronomical library, being the oldest star map known from any civilization, the *Dunhuang Star Chart* from Tang dynasty (7th–8th century) China is a book in the form of a scroll. It is named after the location of its rediscovery, in a cache of manuscripts preserved for centuries in a desert cave near Dunhuang, in northwestern China.

Along with the chart (see page 18), the cache contained a few other astronomy related documents with details of Chinese calendars and almanacs, although most of the documents were religious texts about Buddhism. The astronomical documents record the ancient Chinese constellations from three different schools of astronomy (led by the astronomers Gan De, Shi Shen and Wu Xian, respectively), and also provide details of how the lunar cycle lined up with the seasons. One of the most widely known cultural calendars from China, the Chinese zodiac, in which each year is associated with a particular animal on a 12-year cycle is also discussed. This 12-year cycle comes from tracking the motion of the planet Jupiter across the skies. Jupiter has an orbital period of 11.86 years. In ancient China this was rounded to 12 years, and observations of Jupiter passing through each of 12 'Jupiter stations' in the sky in a repeated cycle set up the 12-year Chinese zodiac. So Jupiter is known as *Suixing* in Chinese – 'the year star'.

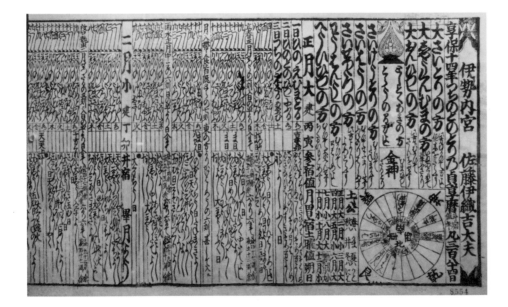

LEFT

Jokyo Calendar

A version of the Jokyo Calendar dating from 1729. It is on display in the National Museum of Nature and Science, Tokyo, Japan.

Jokyo Calendar

Not quite a book, but close, this document from 1684 records innovations to the astronomical lunisolar calendar used in Edo-period Japan, which were made by the astronomer Shibukawa Shunkai. Shunkai refined the length of the solar year (to 365.2417 days).

Introduction to the Study of Astronomy

Written in the early 18th century by the Japanese intellectual Baba Nobutake (?–1715), this book, first published in 1706, summarizes astronomical knowledge of Edo-period Japan. The author was a prolific writer, producing over 30 books on topics which ranged from divination practices in Japan to historic military tales, magic tricks and astronomy. In this book he discussed solar eclipses, a magnetic explanation for the tides and presented the Tychonic model for the universe (hybrid geo/helio-centric – see Chapter 4: Developing Our Model of the Universe).

The Edo period was a time of isolationism in Japan, when military leaders were concerned about the impact of European culture on Japanese traditions. The import of European books, including ones about astronomy and other scientific advances, was prohibited. Around the same time Nobutake published his astronomy book, the astronomer Nishikawa Joken (1648–1724) was interested in teaching the astronomical advances that were happening in Europe. In 1712 he published *Tenmon Giron* (A Discussion of Astronomy), and in 1720 he gave a lecture about it to the country's military leaders, which is thought to have played a role in convincing them to relax the rules on importing books.

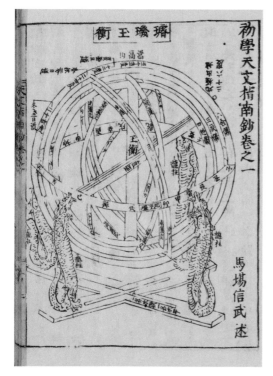
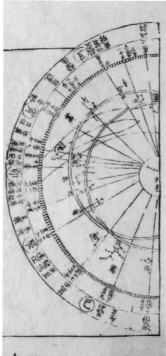
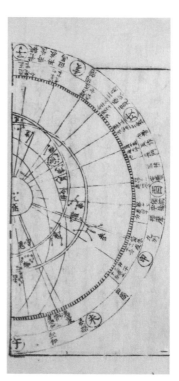

DEVELOPING OUR MODEL OF THE UNIVERSE

How do we make sense of the universe we live in? Humans have wondered about the heavens above them since before recorded history, and as we noted in Chapter 3, astronomy has impacted cultures across the world in fascinating and diverse ways. Astronomy can be considered to be both one of the most modern and the most ancient of the sciences. In this part of the Astronomers' Library, we'll collect some of the books that record our developing understanding of the universe we live in.

Our everyday experience suggests we live on a flat Earth, with the Sun, Moon, planets and fixed stars rotating overhead in a 'celestial sphere'. However, look a little more carefully and evidence emerges that Earth must be a globe. This was worked out by Aristotle (see page 150–151) and it is clear from medieval astronomical teaching texts (see Sacrobosco, pages 109 and 228) that the roughly spherical shape of the Earth was well understood by that time.

The geocentric (or Earth-centred) model was a little harder to topple. It really looks like the skies circle us, and even accurate predictions of the motions of the Sun, Moon, planets and stars are possible in a geocentric model (although the details start to get a bit complex – with things like epicycles, and lots of corrections needed to make very accurate predictions). A scientific revolution, set in motion in the 1500s by Copernicus (see pages 164–167) and given fuel by astronomers like Kepler (see pages 168–171) and Galileo (see pages 176–183) in the 1600s, was needed to convince humanity of the extraordinary fact that Earth moved around the Sun, and it took decades for this heliocentric, or Sun-centred, model to be accepted as the ground truth.

But the universe doesn't stop with the Solar System, and the development of both better telescopes and improved mathematical techniques opened the eye of astronomers to both new planets (with the discovery of Uranus in 1781 and Neptune in 1846) as well as a sea of stars. No longer were stars dots of light on a fixed 'celestial sphere', but enormous spheres of light just like our own Sun, extending to vast distances. Careful counting of the density of stars in the sky revealed we live in a structure of stars, which

we see as the Milky Way stretching across the darkest night skies. In 1785 the brother-sister team of German-English astronomers, Caroline and William Herschel (1750–1848 and 1738–1822, respectively) published the first attempt to map the vast 'collection of stars' around us – although that model suggests a galaxy much smaller than it really is and kept the Sun at the centre (which it isn't). The first measurements of distances to stars (around 1832) revealed just how distant they were, and started to set the scale for understanding the enormity of the universe.

Thanks to improved telescopes, astronomers were also starting to notice 'spiral nebulae.' While records existed of these objects in the earliest existing star charts (e.g. Andromeda is mentioned as a 'little cloud' in *The Book of the Images of the Fixed Stars*, from 964 CE – see page 22), the first clear drawings we have would have to wait until the development of very large

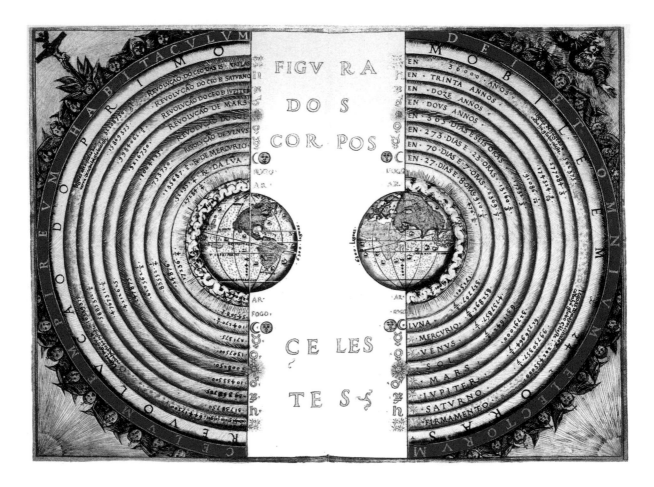

telescopes (e.g. the Whirlpool Galaxy through the 'Leviathan of Parsonstown' telescope from 1845). Debate was soon underway as to their nature (e.g. in *Other Worlds Than Ours* by Proctor in 1878 – see pages 92–95). We know today that these are collections of stars like our own galaxy, just much, much more distant. This debate would only be settled in the 1920s, once Henrietta Leavitt (1868–1921) developed a method using variable stars to estimate distances to external galaxies. Using this method and observations of the almost universal 'redshifts' of spectra[1] of external galaxies from observations by Vesto Slipher (1875–1969), Edwin Hubble (1889–1953) noticed more distant galaxies have larger and larger redshifts, and proposed we must live in an expanding universe of galaxies. Even our Milky Way Galaxy isn't the centre of the universe!

ABOVE AND OPPOSITE
Right and wrong

Medieval European illustration of how to use observations of ships at sea to tell Earth is a globe, from *Sacrobosco* (see pages 109 and 228).

A geometric (Earth-centred) map of the universe from 1568.

---

[1] A spectrum is a kind of detailed rainbow – this is useful because atoms in distant galaxies create light in a distinct pattern in a spectrum which can be measured. The Doppler shift is an effect which moves the colour of this pattern just a little bit either to the red (if the source is moving away) or the blue (if the source is moving towards us). Almost all galaxies have redshifted spectra.

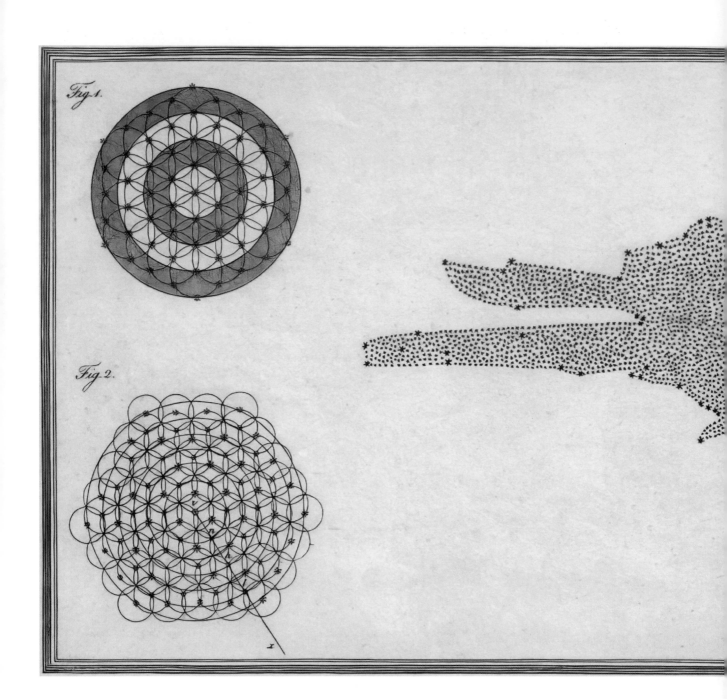

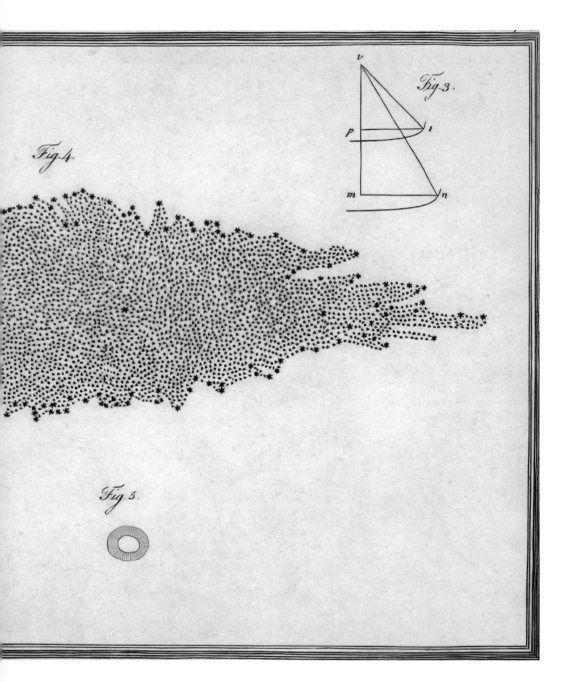

Galactic mapping

The first ever map of our Galaxy, published by William Herschel in the *Philosophical Transactions of the Royal Society of London* in 1785. Almost everything about this is known today to be wrong (it is too small and puts the Sun at the centre), but it marked a huge conceptual leap that we could map the system of stars we live in.

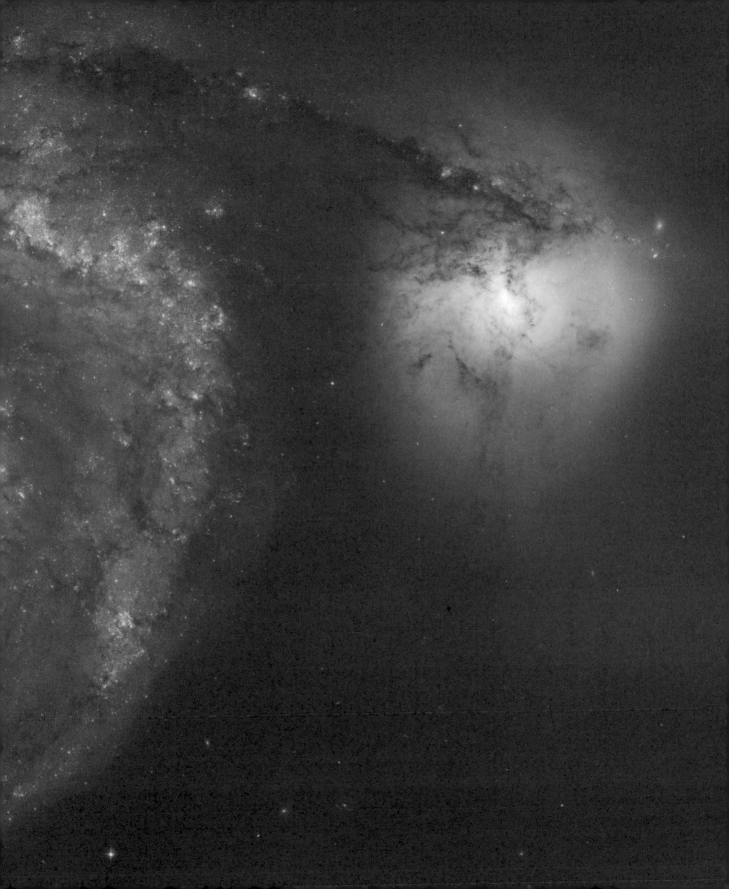

Today we understand that we live in a universe of billions of galaxies, each made of billions of stars – giant spheres of hot gas burning via nuclear fusion in their core, just like our own Sun. Around most stars – perhaps even all – many planets orbit. This vast universe began about 14 billion years ago in an event called the Big Bang, at which time the distances between any two points was zero. Since then, the separation between any two points in the universe has continuously expanded, and today, the best telescopes in the world can look back across billions of light years of cosmic history to galaxies at the very edges of the universe. Observations of the motions of stars in galaxies and galaxies in clusters, interpreted by our understanding of gravity (which started with the *Principia*, see pages 200–205), reveal evidence of mysterious 'dark matter', which adds to the gravitational forces but creates no light. And detailed observations of the expansion rate of the universe now show that it is accelerating, revealing the need for an even more mysterious, 'dark energy' to drive this acceleration. Astronomers today eagerly await better understanding of the natures of these mysterious components, as we seek to continue to improve our model of the universe we live in.

De caelo et mundo

Aristotle (384–322 BCE) was an Ancient Greek philosopher whose work, which covered an amazing array of topics across science, philosophy and culture (e.g. politics), had a significant impact on European scholarship for centuries. Many of his ideas – across several areas of science, including astronomy – were not significantly challenged until the 17th and 18th centuries, meaning his world view dominated scientific thinking for almost 2,000 years.

The main (surviving) work on cosmology by Aristotle, *De caelo et mundo* (On the Heavens), was published around 350 BCE. It lays out the Aristotelian world view of a universe surrounding an imperfect and corruptible Earth, filled with 'perfect' objects, moving along 'perfect' (by which he meant spherical or circular) paths. This was the classic geocentric (Earth-centred) universe, and it also made a clear distinction between things on Earth – made of the four elements (earth, water, air and fire) and subject to decay and destruction – and things in the heavens which are pristine and eternal. This idea would persist in astronomy until well into the 16th century, and beyond.

RIGHT AND BELOW

Aristotlean universe

Diagram showing the natural place of things, the 'Aristotlean concept' described in *De caelo et mundo*.

Diagram showing an Earth-centred universe. Both images are from the 1614 publication *Disquisitiones mathematicae* by Christoph Scheiner.

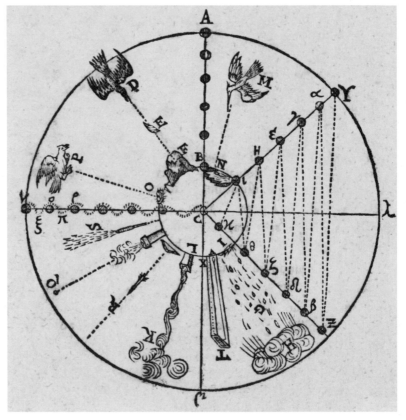

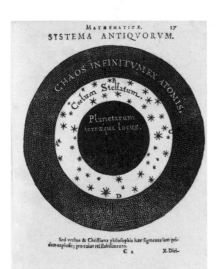

On the Sizes and the Distances of the Sun and Moon

Without the benefit of modern search engines, how would you go about working out the distance to our nearest celestial neighbours, the Moon and the Sun? It turns out that triangles are the answer.

The Greek astronomer and mathematician Aristarchus (310–230 BCE – meaning Aristarchus of Samos – not one of the many other Ancient Greeks named Aristarchus) made a series of geometrical arguments using ancient Greek trigonometry in his only surviving book *On the Sizes and the Distances of the Sun and Moon*. His ingenious methods are based on a guess (a correct one) that the light from the Moon is just reflected sunlight. That means that when the Moon is at its quarter phase (exactly half illuminated), it forms a right-angled triangle with the Sun. By measuring the angle between the Moon and the Sun at this exact moment, Aristarchus could work out the ratio of the distances to the Sun and the Moon: trigonometry in action! Making this measurement by eye is tricky, so the actual number Aristarchus published (the Sun being 19 times further away than the Moon) was quite a lot smaller than what we know today (it is actually 400 times), but one thing it correctly showed was that the Sun was both more distant, and therefore physically larger than the Moon, since they appear to be the same size on the sky, the more distant object must be larger. More information came from observations of lunar eclipses, a phenomena when the Moon passes through Earth's shadow. The timing of such events allowed for more triangles in space, this time allowing Aristarchus to work out that the Moon must be physically smaller than Earth – he reckoned it was about a third of the size. Both of these facts together led Aristarchus to conclude that the Sun had to be quite a lot bigger than Earth, making a geocentric (Earth-centred) model very unlikely in his view.

Aristarchus of Samos is supposed to have argued for a heliocentric universe with 'a fire at the centre of the universe' (the Sun) centuries before Copernicus did, but any written description of that model is lost forever, and the idea didn't take off at all at the time; his only remaining text describes a geocentric – or Earth-centred – universe.

BELOW AND LEFT

On the Sizes and the Distances…

Diagram showing the geometry of the Sun (bottom), Earth (centre) and Moon (top), just as the Moon exits the shadow of Earth – shown by the triangular shape.

A cherub adorns an illuminated 'L' (for lunam) from the title page of a 1572 edition.

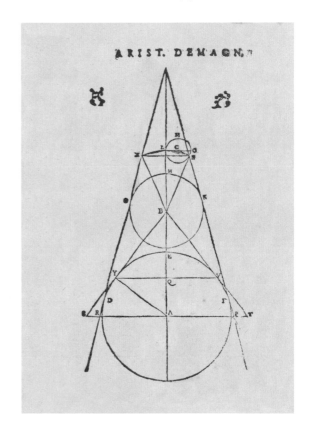

The Almagest

The Almagest by Claudius Ptolemy (c.100–170 CE) is one of the most significant books in the history of astronomy to have presented a geocentric world view. We now know this model to be incorrect (the Sun is at the centre of the Solar System) but nevertheless this model, as described in *The Almagest*, was foundational understanding for astronomers in Europe and the Islamic world for centuries.

The Almagest was first published in Latin, in Egypt, with the title *Mathēmatikē Syntaxis* (Mathematical Treatise), during a highly peaceful and prosperous era of the Roman Empire that is sometimes known as the 'Century of Five Good Emperors'. The book survived to the modern age largely through the Arabic tradition of copying manuscripts, where it became known under the title *The Greatest Treatise*, or in Arabic *Al-majisṭī*, which was then anglicised to *The Almagest*.

Claudius Ptolemy lived in Alexandria in Egypt. It has been suggested he was born in Greece and later took a Latinized name as a Roman citizen, for the Greek version of his name, Ptolemaeus, was common in Ancient Greece. He goes down in history primarily as an astronomer, but he also made contributions to mathematics, geography, optics, musical theory and philosophy.

Ptolemy was not the first classical scholar to discuss a geocentric model of the universe. We already discussed how Aristotle presented his idea of the perfect geocentric universe around 450 years before (see pages 150–151). There is also evidence

RIGHT AND OPPOSITE

The Amalgest

One of Ptolomy's 'Handy Tables'. He produced a number of these for reference to the skies. This, one of the oldest surviving copies, dates back to c.813 CE.

A zodiacal diagram from a 1528 (coloured) edition of *The Almagest*. The outer ring shows illustrations of the mythology of the 12 classical Greek zodiacal constellations (i.e., the constellations the Sun passes through annually).

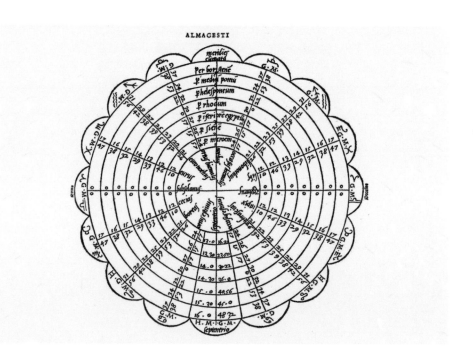

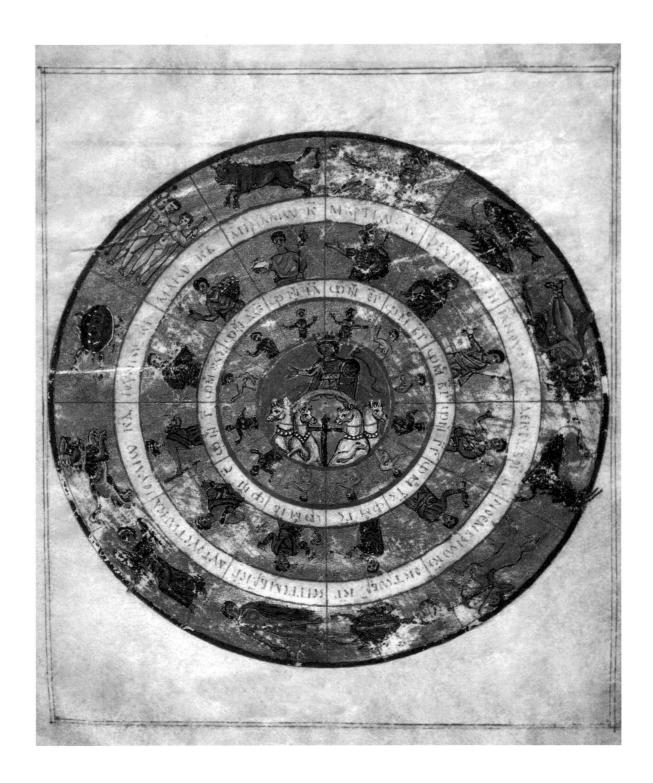

RIGHT

The Amalgest

A list of star positions, from a
translation of *The Almagest*. This
edition was published in 1490.

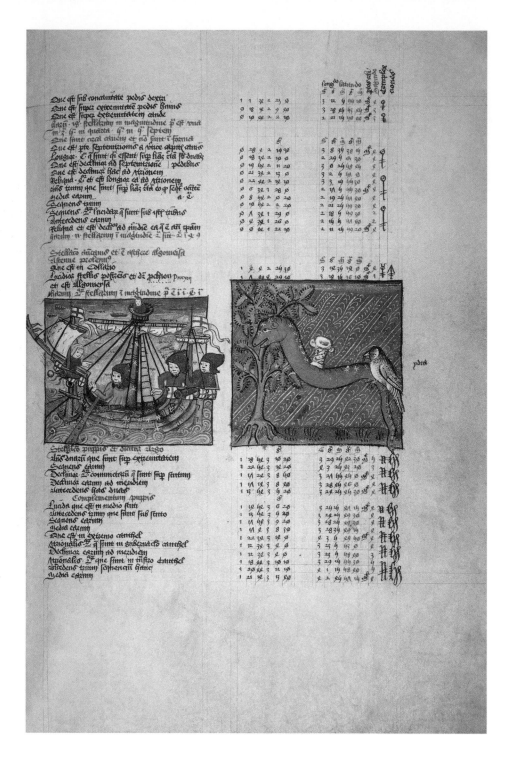

that something like 200 years earlier the Greek astronomer Hipparchus (c.190–120 BCE) had described a physical geocentric model for the skies above Earth which could also be used to predict the motions of the planets.

Hipparchus is also credited with inventing the astronomical magnitude system, as a ranking from 1 to 6 for the brightness of visible stars in the skies. Generations of astronomy students (and professional astronomers) curse the continued use of this system set up by him, since as a reverse numerical (and logarithmic) scale for the brightness of stars it often causes confusion. However, no astronomical texts from Hipparchus remain, so we only know of this via mentions in other publications, including in *The Almagest*.

In *The Almagest*, Ptolemy laid out the classic geocentric model for the universe (at that time consisting of only the Solar System and the fixed stars) and provided details of how to use it to predict the motions of the planets. The main features of the cosmological model Ptolemy adopted were that the universe consisted of a series of nested spheres, with Earth being a fixed sphere at the centre of everything. This model followed Aristotle and Hipparchus before in assuming that the sphere was the perfect shape and therefore the celestial heavens particularly must consist of spheres, and make spherical, or circular, motion; and that obviously Earth does not move. Under this physical model, Ptolemy went on to include a discussion of the daily motions through the skies of the Sun, Moon, stars and planets, and gave details about how to use this to calculate important dates and times – for example, solstices and the length of the year. There was a section on the motion of the Moon, solar and lunar eclipses, as well as a number of chapters focusing on the motions of various planets ('wandering stars') across the background of the fixed stars.

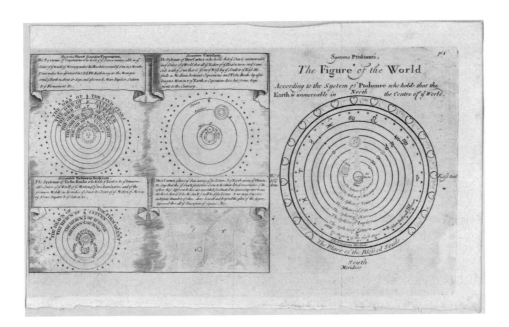

LEFT
'The Figure of the World'

An English-language map created to show the Ptolemaic system of the world.

RIGHT
The Amalgest

A highly decorative
constellation map from a 1515
edition of the book, published
in Liechtenstein.

Ptolemy's book so completely superseded earlier astronomical texts in the classical world, that most only remain in mentions in his and other work. And while no original editions of *The Almagest* remain, translations and copies abound. Most of these show subtle variations, including improvements in the explanations from edits, presumably by other astronomers. Many astronomers would also write commentaries on the book.

One notable edit was made by Hypatia (c.350/70–415 CE), a prominent thinker in Alexandria in the Eastern Roman Empire who lived around 200 years after Ptolemy. Modern scholars now credit Hypatia with improving details and explanations in *The Almagest* for calculating the motion of the Sun in the sky (and possibly others). Her father, Theon, also a scholar in Alexandria, recorded this work in his own commentary on the version of *The Almagest* his daughter had edited.

Hypatia is often recorded in history as the first ever female mathematician. She also lived in Alexandria, Egypt, when it was part of the Eastern Roman Empire. She is recorded to have taught philosophy at her father's school, edited *The Almagest* and was known for her work constructing astrolabes.

Historically, Hypatia is perhaps most famous for her death, at the hands of a Christian mob during a period of unrest in Alexandria. Myths and stories about this event abound in history and fiction, and it has been used as an allegory (or moral story) by multiple groups over the centuries who picked out certain parts or added fictional details. For example, her death is sometimes conflated with the destruction of the Great Library of Alexandria – although the two events were not contemporaneous. We may never know the truth of Hypatia's life and contributions to astronomy, but it is a fascinating story.

Many of the earliest versions of *The Almagest* do not include any large diagrams showing the cosmological model, however, illustrations of the geocentric model abound in early European printed astronomical texts, since European and Islamic astronomers adopted it as the basis of astronomical knowledge for over 1,000 years. The ideas in *The Almagest* would not be seriously challenged until the time of Copernicus (16th century), roughly 1,400 years after its first publication!

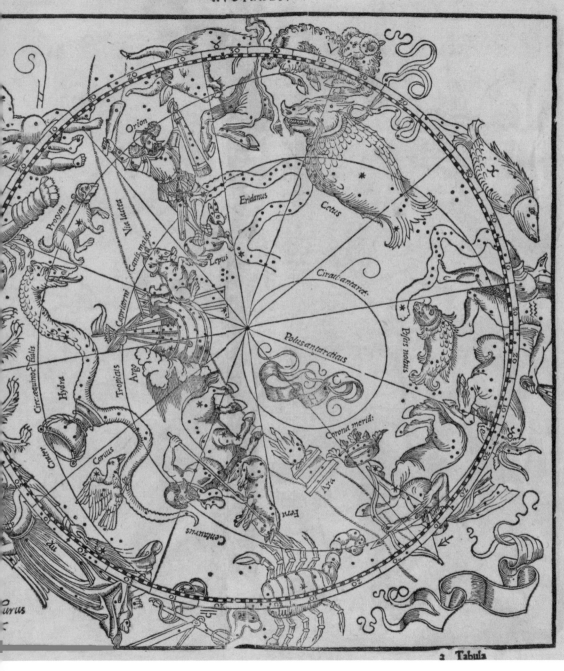

2 Tabula

Astronomicum caesarium

Astronomicum caesarium is often described as the most beautiful astronomy book ever published, and it certainly has a good case to make. Written by Petrus Apianius (1495–1552), it was first published in 1540, in Ingolstadt (now Germany).

In total the book contains 36 separate woodcut illustrations, many of which have hand-painted colour added, and 21 of which are 'volvelles', or moving paper models which can be used as tools to calculate the motions of planets, timing of eclipses and more. It is a stunning book, fit for royalty just as its name implied.

Petrus Apianius was a 16th-century German scholar most famous for his work producing beautifully illustrated astronomical works, including *Astronomicum caesarium* and also the earlier *Cosmographicus liber* (sometimes referred to as *Cosmographia*). He was a contemporary of Copernicus but published a geocentric view of the universe. In his, day his books were extremely popular, and he was a favourite of the Holy Roman emperor Charles V.

Astronomicum caesarium presents a solidly geocentric model for the universe. Its name is usually translated to Emperor's Astronomy, and it was produced under the patronage of two European monarchs: Charles V and his brother, who would follow as Ferdinand I. The primary audience at the time of its publication would likely have been court astrologers wishing to calculate accurate horoscopes for their monarchs, and perhaps the fanciful illustrations on the accurate paper models which could make the precise calculations were just part of the show for their royal patrons. Approximately 100 copies were made and sent to many prominent European monarchs, including King Henry VIII in England. A number still exist today, mostly in large libraries.

While it is interesting (and beautiful) astronomically, it is perhaps most noteworthy as a pinnacle of the use of volvelles in bookmaking. These paper wheel models have been used in many types of publication, but perhaps found most use in astronomical texts, where they act as a type of astrolabe. Astrolabes as objects are usually made from concentric metal circles, which can be rotated to make astronomical calculations, while a volvelle is made from a series of concentric paper circles and cut-outs on the page of a book but used for the same, or similar, purpose. Volvelles had first been developed in the 13th century, but of course not many examples remain. It should be fairly obvious that they are both complex and costly to add to books, and also fairly delicate to preserve. Apianius created several books that featured volvelles, but by far the most elaborate and beautiful is the *Astronomicum caesarium*.

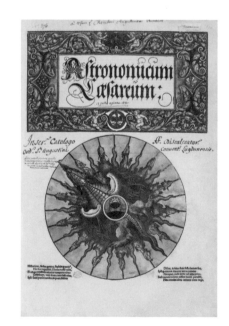

RIGHT

Astronomicum caesarium

Cover of the 1540 edition. The large dragon rotates around the 27 small dragon/head tails, which relate to the motion of the Moon around Earth, with the big dragon showing the two lunar nodes (where the path of the Moon crosses the path of the Sun).

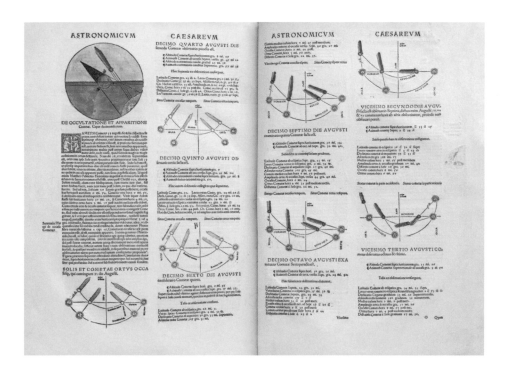

As well as the volvelles,
the book contained several
observations of comets,
noting (correctly) how their
tails always point away from
the Sun.

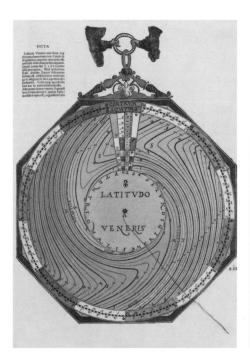

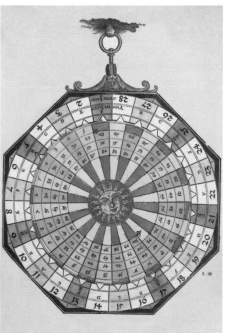

ASTRONOMICVM

ENVNCTIATVM VICESIMVM SEPTIM.

Perspecta iam Lunæ defectus possibilitate, vt sic loquar, id est, tempore quo possit contingere, per instrumentum enunctiati vicesimtertii. Vlterius nunc vtrum scilicet nec certò fiat, quantáq; eclipsis futura sit præsenti figura perspicere.

Ppositionem veram diei, qno futuram Lunæ obscurationem præuideras, possibilem saltem, simulatæ perspexeris, Argumenta Solis & Lunæ ad diem eundem per enunctiata 12 & 25 comporta. Verum quoq; Lunæ loci, eiusdem latitudinem, Latitudinisq; argumentum, necnon capitis draconis lo cum verum per enunctiata 17 & 18 congere. His enim congestis, ea quæ ad eclipses attinent, cuncta contuleris. Per instrumentū q̃q enūct; 18 eclipsis possibilitas, tam Solis quàm Lunæ haberi potest, & melius q̃ in priori, hoc modo: Vbi indicem capitis draconis rite locaueris, ad tempus sc̃ veræ oppositionis vel coniunctionis, locum quoq; Lunæ in Zodiaco contueberis, per quem filum directù si inter d & a (Luna iuxta caput draconis morante) vel inter g & e (Luna caudæ draconis vicina) ceciderit, fateri aude Lunam in illa oppositione defecturam. Vicissim Solis defectum conspicaberis in o̅ne quadā hac lege. Tempora o̅nis verŭ cum filo signa, quod si inter b & e iuxta caput draconis, vel inter f & g iuxta caudam draconis feratur, contingens est Solis deliquium in hac coniunctione, Verũ Solis obscurationes propter aspectuum diuersitates non ita ad vnguem, sine calculandi opera, sicut Lunæ haberi nobis potuerunt, eas tamen, deo volente, in posterum quoq; in instrumenti formam redacturos nos, quo ad licebit, nõ diffidimus. Possibilitate eclipsis Lunaris agnita, lineam in plano aliquo præscribe, quam H D nomina, cuius medium A sit. Huic alia perpendicularis inducta B F dicatur. Pedem nunc circini alterum in A litera fige, extensus alter circulum occultum pro Arbitrio scribat, qui idem in linea B F & punctis B F secetur. Eo facto Lunæ latitudinem considera, quæ si Septentrionalis fuerit, tum oportebit Argumentum eius verum esse 5 Signorum & 18 graduum ad minus, vsq; in 6 Signa etiam, vel Signi nullius, graduumq; 12. Iam si Argumentum sit Signi nullius, diuidatur quarta B D in 18 partes sic. primo in tres, deinde singulas in alias tres, postea quamlibet in duas, & habebis 18 portiones, quarum singulæ 5 gradus continent. Proximo B literæ puncto C ascribe, rectamq; lineam à centro A per C educito. Si vero Argumentum latitudinis Signorum sit 5, tum B H quartam dissece, primæ autem post B literam sectioni, M literam appone, perq; M & A centrum lineam, vt antea, diduc, Quod si meridionalem Lunam conspexeris, Argumentum habens 6 Signorum, & insuper aliquot graduum, tum ab F litera D versus proximum punctum elige, eundem cum T litera signans. Argumento 11 Signorum existente, ab F versus H cum puncto huiusmodi digredere, cui ascribe G. Hæc in partem mouearis quicquàm, si gradus aliquando Signis non adiunctos videas, obseruatis diligenter Signis tantummodo.

Hæc omniũ &c.

CAESAREVM

Hæc omnium præsens figura iterat, posterius DISPOSITORIVM dicenda.

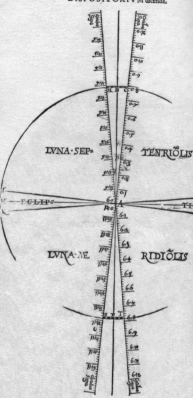

LVNA·SEP·   TENRIOLIS

ECLIPS·   TICA

LVNA·ME   RIDIOLIS

Deinde in limbo exteriori sequentis figuræ, quæ post hac, instrumentum Semidiametrale vocabitur, Solis Argumentum vestiga, cui filum vbi superposueris, nota eiusdem, & limbi interioris (quæ vmbrarum varietates per secunda metitur) contactum, secunda ve ro illa à filo abscissa, seorsum excipe. Argumentum deinde Lunæ perquirens, cum filo pariter transmisso signa, quod filum binas tibi corporis vicz Lunaris & vmbræ terræ, semidiametros aperit. Filo sic durante, circinum in centro fige, pede altero in punctũ vsq, in quo nigrior area à filo tangitur, extenso, illa enim extensio semidiameter corporis Lunæ est, quam ita inuariatam reserua. Alium post hac circinum fini prioris diametri iam habiti infige, Pedem alterum eiusdem circini vsq ad areæ fuscæ terminum protendens, semidiametrum quoq vmbræ terrenæ habebis. Aream hanc circum ferentia quadam candenti circuncirca includi vides, vltra quã aliam, calami latitudine, superficiem cernis, quæ in 60 secunda distributa est. Sed cum iam antea secunda varietatæ vmbræ didiceris, pedes

des circini pro tot secundorum distantia contrahi debent, & vera vmbræ semidiameter habebitur. Talem postea circinum in A plani prius ad hoc præparati, vbi firmaueris, circulumq; descripseris, veram vmbræ quantitatem cernis, vmbræ inquam terræ, quæ in Lunæ transitu, oppositionis illius tempore fit. Mox in instrumento cui nomen dispositorio, Lunæ Argumentum in lineis, siue A C siue A M, siue A G, siue A T sit, contemplator. Loco oblæ to pedem circini insere, extendasq alterum in argumenti punctum. Eandem postea circini extensionem parato plano infer, pedeq altero in A fixo, cum altero punctum in linea paulo antea, huic rei deputata, exprime, literæq M signa. Iam circinum require, extensum, iuxta corporis Lunaris semidiametrum, quem vbi fixeris in M, cumq eodem circulum descripseris, corpus Lunæ te habere potes. Ex illis nunc demum certus esse potes, de contingenti huius oppositionis eclipsi, deq quantitate eiusdem. Si enim corpus Lunæ totum sub vmbram concesserit, vniuersalis, si partim, particularis, sin omnino non, nulla eclipsis euenit.

The volvelle version of
the cover image dragon,
which can be rotated to
find the position of the
lunar nodes at any date.

A lovely illustration of
star constellations.

De revolutionibus orbium coelestium

Even in the set of books we cover in the Astronomers' Library, not many can lay claim to having set off scientific revolutions, but *De revolutionibus orbium coelestium* (literally, On the Revolutions of the Heavenly Spheres), sometimes nicknamed simply *De revolutionibus*, certainly can. It was written by Nicolaus Copernicus (1473–1543) and was first published in 1543, in Nuremberg.

Copernicus was born the youngest of four siblings in a part of Europe which is now located in Poland (in 1473 it was part of 'Royal Prussia'). He was highly educated, across what to modern ears is an astonishing range of disciplines, from classics and languages to religion, mathematics and medicine. He made important contributions to the development of economic theories of his time as well as to astronomy. He lived to the age of 70; his death came after a period of ill health in which he was persuaded to publish his life's work – just in time (legend has it) to see the final, printed copy of the publication on the day he died.

In this book, Copernicus explained why it should be the case that the Sun was at the centre of the universe, instead of, as had been previously believed, Earth. He argued for a model in which all the planets must revolve around the Sun along the surface of perfect spheres, transcribing giant circles in space. In the illustration opposite, we see a diagram of that model, with the Sun at the centre surrounded by perfect circular orbits of the five known planets at the time, Mercury, Venus, Earth and its Moon, Mars, Jupiter and Saturn. The whole system is surrounded by the words *Stellarium fixarum spherae imobillus*, literally a 'fixed sphere of immovable stars', sometimes called the celestial sphere, which represents the rest of the universe. This book was the culmination of a life's work – Copernicus worked on the arguments for over 30 years and was reluctant for many years to release his vision to the wider world, worrying about its reception in an era when many scholars thought the idea of Earth itself moving was laughable and contrary to their religious beliefs. In the end he was persuaded to publish but he only just lived to see it happen.

The six distinct sections (sometimes called separate books) in *De revolutionibus* lay out both the philosophical and mathematical arguments linked to Copernicus's ideas of a 'heliocentric' (or Sun-centred) model for the universe. Curiously, because his book came before the development of modern ideas of the scientific method, it provides nothing we would today call scientific proof that the heliocentric model was correct. In fact, the Copernican model with his insistence on perfectly circular orbits for all the planets was in some ways worse at predicting the motions of the planets than the geocentric models it argued to overturn. Instead, the main argument Copernicus made was that it made sense that the Sun should be at the centre of the Solar System – that it made a more complete and elegant

solution. The book also contained long, highly mathematical sections on spherical trigonometry, and details of the orbital motions of Earth, the Moon and other planets. As a result, it known in certain circles as 'The Book No One Read', although in his similarly titled work on the book's readership history, the astronomical historian Owen Gingerich (1930–2023), who personally inspected 600 existing printed copies of the book for information in annotations by readers, disputes this idea.

Of course, some people must have read at least parts of *De revolutionibus* since it started a scientific revolution that we still talk about today. The pace of the spread of scientific ideas at that time was slow, so it took literally decades for the ideas to be generally accepted, and it wasn't until later tweaks to the model (removing the requirement for circular orbits) that it started to actually make predicting the motions of planets easier and more accurate than the geocentric models it displaced. As such, astronomical books and models sometimes present a geocentric viewpoint of the universe even many decades after *De revolutionibus*, despite ultimately Copernicus having been right about the Sun being at the centre of at least the Solar System so long before they were written.

The main revolution of *De revolutionibus* was the demotion of Earth from the centre of the known universe to just one of several planets orbiting the Sun. As a residual of this, even today a 'Copernican' principle in cosmology is an idea which states that we don't observe the universe from any particularly special place in it; for example, that the laws of physics should work the same everywhere.

Harmonia macrocosmia

The Copernican Solar
System, illustrated in 1661
by Andreas Cellarius. See
pages 35–40 and 188–190
for more from this book.

1. Kepler's Effect

Not many astronomers have contributed more than one book in our Astronomers' Library; from Kepler we collect three. His contributions to the development of our understanding of the shape of the Solar System as well as the laws of motions of the planets were so significant they are still taught in all introductory astronomy courses today. However, reading his books reveals a fascination with astrology and the geometric and musical harmonies of the universe, which is much less discussed.

Johannes Kepler (1571–1630) was a notable 17th-century astronomer and astrologer who was born in Germany and lived there as well as in Austria and Prague. He spent some years working with the observational astronomer Tycho Brahe (see page 172), and it is his careful star charts tracking the motion of the planets in our skies that Kepler would use to develop his laws of planetary motion through space. In addition to his work on the motions of planets, Kepler recorded the observation of a supernova in 1604, now known as 'Kepler's Supernova' and which we today observe as a nebula supernova remnant.

Mysterium cosmographicum

Mysterium cosmographicum was the first published book laying out support of the Copernican heliocentric model of the universe. In it, a young Johannes Kepler – just 26 when this book was published in 1597 – described a universe in which planetary orbits for the six planets known at this time are separated by distances set by the five 'Platonic' solids. A Platonic solid is a three-dimensional shape in which all sides are identical. They are the tetrahedron (four triangular sides), cube (four square sides), octahedron (eight triangles), dodecahedron (12 pentagons) and icosahedron (20 triangles). In Kepler's model universe, each planetary orbit was described by a sphere, which would be nested within each other, with the Sun at the centre, and separated by each of these five shapes in turn.

The model actually works reasonably well to predict the pattern of distances to the planets. Each of these five solids has something called an inradius – the radius of a sphere which can just fit inside the shape – and a circumradius, or the radius of a sphere which can just enclose the shape. No matter the size of the shape, the ratio of the radius of these two spheres is the same, and you can pick one of each of these shapes to correspond reasonably well to the radio of the radius between adjacent planetary orbits of the six bright planets (Mercury, Venus, Earth, Mars, Jupiter and Saturn). For example, Kepler places a cube between the sphere of Saturn (his outer sphere) and of Jupiter. The ratio of the orbital radii of those two planets is 1.8 (Saturn orbits at a distance 1.8 times further from the Sun than Jupiter is). For a cube, the circumradius is 1.7 times the inradius (you can fit a cube into a sphere which is 1.7 times bigger than the biggest sphere you can fit inside the same cube). So, as you can see the match is OK although not perfect. Kepler was certainly impressed.

The full title of this book is *Prodromus dissertationum cosmographicarum, continens mysterium cosmographicum, de admirabili proportione orbium coelestium, de que causis*

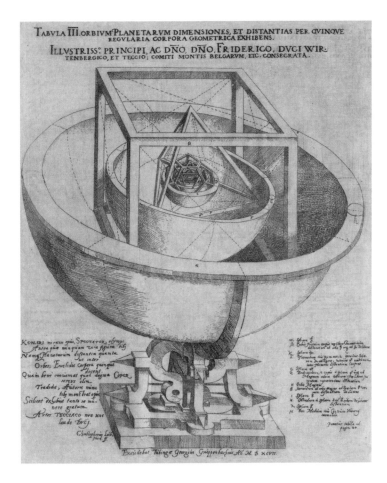

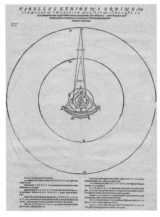

coelorum numeri, magnitudinis, motuumque periodicorum genuinis & proprijs, demonstratum, per quinque regularia corpora geometrica which translates to show just how excited Kepler was by this model: Forerunner of the Cosmological Essays, Which Contains the Secret of the Universe; on the Marvellous Proportion of the Celestial Spheres, and on the True and Particular Causes of the Number, Magnitude, and Periodic Motions of the Heavens; Established by Means of the Five Regular Geometric Solids. He wrote about how he had revealed the 'secret of the universe' in 'God's geometrical plan'. While much of what is presented here is now ignored, this book was the first step in his work to refine the simple Copernican heliocentric model to something very close to what we still use today to describe our Solar System. Not least because in response to this book, Tycho Brahe wrote to Kepler to state that only observations of the motions of the planets would truly reveal the secrets of the universe. This communication is said to have inspired Kepler to seek out Tycho Brahe to collaborate – and the rest is history.

ABOVE
Platonic solids

A selection of illustrated pages from *Mysterium Cosmographicum*, including one demonstrating (left) his theory about patterns in planetary separations being due to platonic solids and (right) various diagrams illustrating his famous empirical rules about the orbits of planets.

Astronomia nova

Astronomia nova (New Astronomy) was the book that came out in 1609 in which Kepler, by then in his mid-thirties, published the first two of his three laws of planetary motion. By this time Tycho Brahe had shared with Kepler his observations of the motion of Mars in the skies, showing very clearly its repeated retrograde – or 'backwards relative to the fixed stars' – motions. Kepler's revolutionary idea presented in this book was that planets (including Earth) moved on 'orbits', or paths, in space (rather than being stuck to fixed spheres). He struggled for a while to find a specific shape of path which would allow him to describe the motion of Mars as seen from Earth, but he eventually came up with the idea of a simple ellipse. His first law thus became 'all planets move in ellipses, with the Sun at one focus', an idea which was built on in his demonstration of how that worked for Mars and Earth in this book.

This new model meant that planets have varying distances to the Sun. Their closest approach is called a perihelion, while the most distant point is the aphelion. Kepler had noticed how both Mars and Earth moved more quickly when they were near perihelion, and more slowly at aphelion. He explained this as due to what he believed was the 'motive power of the Sun' weakening with distance, which while not expressed in modern language is actually pretty close to how we conceptualize Newtonian gravity today. He was able to describe the pattern seen in the observations of Mars and Earth with the simple geometrical rule 'planets sweep out equal areas in equal times', which became his second law of planetary motion.

Tycho Brahe had passed away in 1601, and his heirs disputed Kepler's right to publish this work since it was based on Brahe's observations. It is recorded that while Kepler finished this book in 1604, it was not published until 1609 as a result of these legal disputes.

BELOW

Astronomia nova

Illustrations of various geometric calculations of the motion of planets including (left) a comparison of the methods in the models of Copernicus (heliocentric), Ptolemy (geocentric) and Tycho Brahe (who presented a hybrid model).

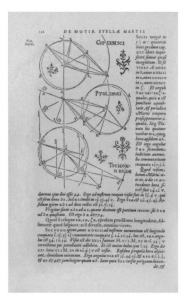

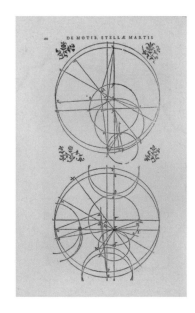

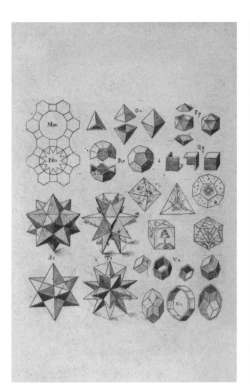

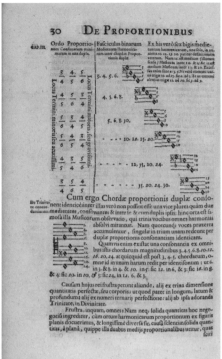

Harmonice mundi

Kepler would complete the publication of his three laws of motion in *Harmonice mundi* (literally, The Harmony of the World) in 1619, ten years after the publication of the first two laws. In this book, Kepler mostly focused on explaining the size and shape of objects in the natural world (including the orbits of the planets) in terms of musical harmonies. Kepler was not the first scholar to consider 'the music of the heavens'. Ideas along these lines had been considered by several scholars from antiquity. However, in this book Kepler would lay out how he believed that musical harmonies set the scale of the Solar System, alongside geometric rules, music, astrology and meteorology. It is interesting to notice that in a book in which Kepler lays out his highly mathematical third law are also ideas about how the musical harmonies between the souls of heavenly objects and those of humans might explain why astrology should work.

Kepler's third law, published in the last chapter of this book, states that 'the square of the period of the orbit increased with the cube of the mean distance to the planet', and this rule genuinely describes what we still see in the Solar System. It is understood today as a consequence of Newton's laws of gravity, and how gravity and the rules of inertia determine how quickly planets move in their orbits. However, Kepler's ideas about the souls of heavenly bodies are no longer considered part of astronomy.

De mundi aetherei recentioribus phaenomenis liber secundus

Tycho Brahe (1546–1601) has to be described as one of the most flamboyant and unusual astronomers in history. Born into a noble family in Denmark, he need not have worked at all, but his detailed and careful observations of the night skies instead helped feed a scientific revolution. He seems to have been quite a character: while studying medicine, including medical astrology, in Germany he lost part of his nose in a duel, thereafter wearing a metal prosthetic. After returning to Denmark, he gained favour with the Danish court, who agreed to fund the building of Europe's first professional astronomical observatory for him. The Uraniborg (or 'Castle of Urania' – Urania being the Greek muse of astronomy), was built around 1580 (see page 87 and overleaf) on an island in what is now Sweden (then it was Denmark), although it was sadly destroyed in 1601 after Brahe fell out of favour with the Danish king. Brahe ended his life in exile in Prague, still working on astronomy, now with the assistance of Johannes Kepler in what would be a highly productive collaboration. He died suddenly aged only 54, and it has been speculated that he was murdered by mercury poisoning, although evidence remains inconclusive.

We collect here the second, of what was to be a three-volume series of books, in which Tycho Brahe published his observations of the night skies and laid out his hybrid geo-heliocentric model for the universe. This second volume was the only one published while Brahe was alive; with an initial distribution in 1588. The first volume in the series was incomplete at the time of his death and Brahe's assistant, Johannes Kepler, published the work not long after Brahe died in 1601. The third volume was never completed.

During his life, Brahe was well aware of the Copernican heliocentric model for the Solar System; however, his writings explain that he just could not conceive of a moving Earth. He believed that Earth was too massive to move, and also that since scripture said it didn't move, it couldn't move. He correctly understood that if Earth did move, astronomers should observe some annual backwards and forwards motion of stars (an effect we call parallax), as the closer ones appear to move on the background of more distant ones. This effect is similar to the view of nearby trees moving against the horizon as you drive along a road, or how you can make your finger appear to move against a background by looking at it with one eye and then the other. Even with his careful observational techniques, Brahe could observe no such motion in any stars, so he concluded this lack of parallax meant Earth must be fixed in place, since otherwise stars would have to be unimaginably distant. Of course, today we know that Earth does move around the Sun, and even the nearest stars are so distant that their parallax is too tiny for pre-telescopic observations to measure.

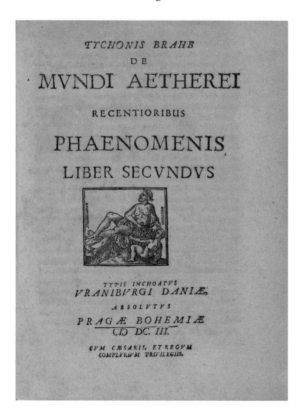

TYCHONIS BRAHE
DE
· MVNDI AETHEREI
RECENTIORIBUS
PHAENOMENIS,
LIBER SECVNDVS

TYPIS INCHOATVS
VRANIBVRGI DANIÆ,
ABSOLVTVS
PRAGÆ BOHEMIÆ
CIↃ IↃC. III.
CVM CÆSARIS, ET REGVM
COMPLVRIVM PRIVILEGIIS.

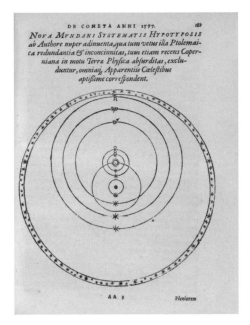

NOVA MVNDANI SYSTEMATIS HYPOTYPOSIS

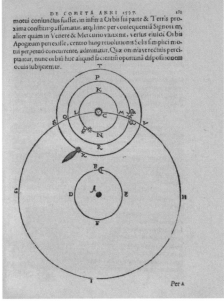

DE COMETA ANNI 1577 diagram

FAR LEFT, LEFT, BELOW

De mundi aetherei...

The Tychonic system, as published in *De mundi*, showing Earth at the centre, with the Moon and Sun circling it, while the other planets (at the time Mercury, Venus, Mars, Jupiter and Saturn) circle the Sun.

A segment of the inner part of the Tychonic Solar System with an observation of the Great Comet of 1577 added along with its supposed circular path around the Sun. Brahe used observations of this comet to conclude they were not atmospheric phenomena (at the time an ongoing debate). However, this comet was certainly not on a circular path around the Sun, and in fact has not revisited the inner Solar System since 1577, so is classed as a 'non-periodic' comet.

A portrait of Tycho Brahe, dating from around the 17th century.

Anyway, based on his observations, Brahe believed he needed a model with a stationary Earth, and, like Copernicus, he was also adamant on spherical motion – following Aristotelian beliefs about the circle being the perfect shape. However, he also clearly saw some benefit to a heliocentric system, so he developed a hybrid model, in which Earth was fixed at the ultimate centre. In Brahe's model, the Moon and Sun orbit Earth directly, while all the other planets orbit the Sun. This model could be used to predict the motions of the planets reasonably well, although to a modern eye the crossing of the orbits of Mars and Earth which it required looks awkward (and dangerous).

This 'Tychonic' system was popular at the time, particularly after the Catholic Church declared the full heliocentric model to be heretic. However, today Brahe is most famous to astronomers for his detailed observations of the skies which enabled Kepler to refine the heliocentric model to the extent that it is basically what we still use today.

Brahe's *De mundi* records some of those data in full page tables, paying special attention to observations of the comet of 1577, whose position is illustrated in a zoom-in diagram of the inner part of the Tychonic Solar System, showing just the Sun, with Mercury, Venus and the comet circling it, all going around the Earth–Moon system.

A portrait of Tycho Brahe

Observatory

Layout of Brahe's observatory
Uraniborg from Joan Blaue's
enormous publication of
1662–1672, *Atlas maior*.

ARCIS VRANIBVRGI,   A TYCHO
IN INSVLA HELLESPONTI DANICI HVENNA CONSTRVCTÆ,   QUO AD TO

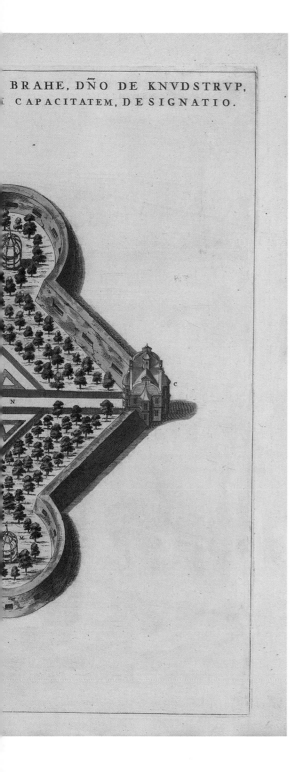

BRAHE, DÑO DE KNVDSTRVP,
CAPACITATEM, DESIGNATIO.

SUN

MERCURY

VENUS

EARTH

MOON

MARS

JUPITER

SATURN

Solar symbols

In many of the maps and diagrams of this book, you will find these traditional symbols, which are used to represent the Sun, Moon and the pre-telescopic planets of the Solar System.

Newer symbols were invented for later discoveries (e.g. Neptune, Uranus and various dwarf planets), however, today the use of such graphical annotation is discouraged by the International Astronomical Union.

Sidereus nuncius

The invention of the telescope by Dutch spectacle-makers around 1608 would revolutionize the astronomers view of the night skies forever more, as well as open up worlds beyond Earth in new and exciting ways.

Galileo Galilei (1564–1642) was the first astronomer to turn a telescope to the night skies and report on his results. His book, first published in Venice in 1610, reports on those observations. The name *Sidereus nuncius* translates as either The Starry Messenger or The Sidereal Messenger ('sidereal' is an adjective meaning 'to do with the stars'). Galileo wrote this short booklet to report on his telescopic observations, and in it he also argued that what he saw was clear evidence for the Copernican model of the universe. He would add other short publications to the *Sidereus nuncius* proper, which are often today bound together, including his *Discorso delle comete* (Discourse on Comets), and *Istoria e dimostrazioni intorno alle Macchie solari* (Letters on Sunspots).

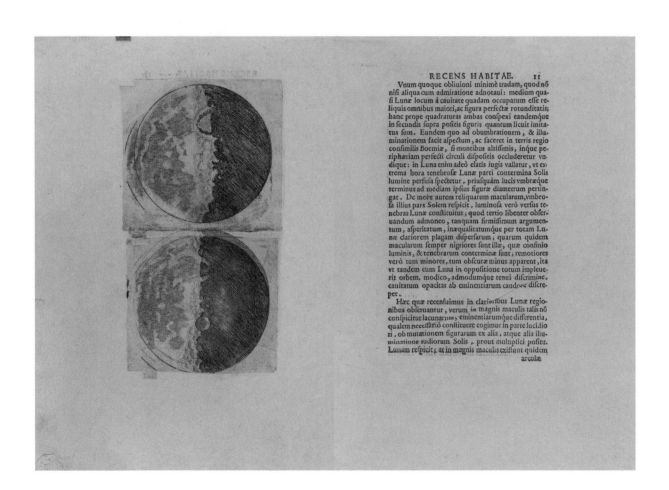

Sidereus nuncius alone contains tens of detailed drawings of the Moon, views of several constellations and a series of observations of what Galileo at the time would call the 'Medicean stars' of Jupiter, which today are referred to as the 'Galilean moons' of Jupiter. The publication of these observations caused a sensation. Galileo described mountains on the Moon – previously believed since the time of Aristotle to be a pristine celestial object in a perfect sphere. Galileo instead described a world people might imagine walking on. Later, in his *Letters on Sunspots*, Galileo would add to this observations of spots on the Sun. Perhaps not a world to imagine walking on, but clearly not a perfect celestial orb either.

Galileo's observations of stars showed that in places on the sky, 'clouds' of nebulosity could be resolved into separate stars through a telescope, suggesting stars which are much more distant than others. Perhaps the universe was much bigger than previously imagined. Galileo's discovery of the 'Medicean stars' of Jupiter and his report described how they moved with Jupiter in the sky, appearing always in a straight line, but to circle

BELOW
Sidereus nuncius

Star clusters as mapped by Galileo in the book: the right page shows the Pleiades, or Seven Sisters, a well-known cluster that is visible to the unaided eye.

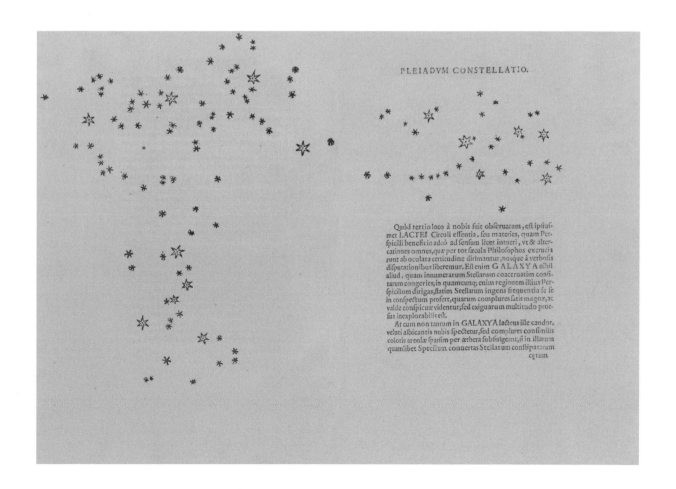

Jupiter over time, so that they noticeably move even hour to hour. This suggestion of objects orbiting Jupiter – not Earth – was to Galileo clear evidence that the models of a geocentric universe was not possible. Clearly not everything orbited Earth.

For Galileo, the final piece of evidence that the Solar System must be heliocentric was his observations of the phases of Venus. Reported in his *Letters on Sunspots*, the existence of these phases meant that sometimes Venus is between Earth and the Sun, and sometimes it is on the other side of the Sun from Earth. Venus must circle the Sun: for Galileo the final evidence he needed.

This book, and the observations and ideas published within it, would bring Galileo to the attention of the Catholic Inquisition. By early 1616 the Catholic Church had decided to declare heliocentricism to be heretical, and ordered Galileo 'to abandon completely… the opinion that the Sun stands still at the centre of the world and the Earth moves, and henceforth not to hold, teach, or defend it in any way whatever, either orally or in writing.' Despite the clear scientific evidence Galileo had found to suggest a heliocentric model was the true description of the Solar System, Galileo agreed to this order, and kept this promise for over a decade.

BELOW
A flare for observation

Drawings of sunspots as published by Galileo in his book *Letters on Sunspots*, 1613.

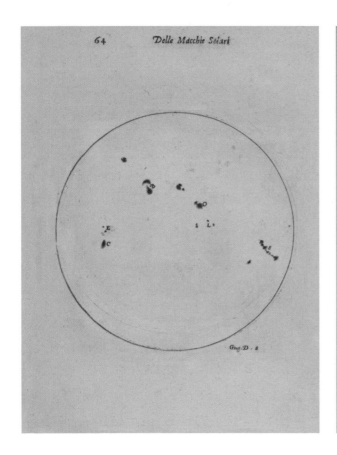

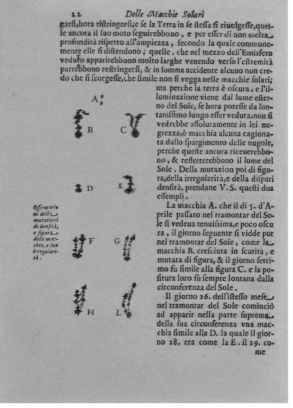

The Medicean stars of Jupiter

This pages shows a series of dated and timed observations of the 'Medicean stars', today known as the Galilean moons, of Jupiter as they circle Jupiter (shown as the larger circle in centre).

Dialogo sopra i due massimi sistemi del mondo

Galileo Galilei made an astonishing array of contributions not only to astronomy and physics, but also to science in general. Many call him the father of modern science – his works were among the first experiments which used something approaching what today we call the scientific method. As well as his famous astronomical observations with the first telescope, he performed many experiments on mass, gravity and velocity, including the famous (but likely apocryphal) experiment of dropping balls from the leaning tower of Pisa.

In the history of books which got astronomers in trouble, this one is notorious. It was published originally in 1632 in Galileo's native Italian, and it almost immediately resulted in Galileo being summoned to the Inquisition, where he was found guilty of breaking a promise he had made 16 years earlier to avoid holding 'heretical opinions' about heliocentrism. Following his trial on the book, he lived under house arrest in Florence for a decade before his death in 1642 at the age of 77.

Unusually for science books, the 'Dialogue Concerning the Two Chief World Systems' is constructed like a script for a play, in which two philosophers and an educated member of the public spend four days discussing the different possible models for the Solar System (the Copernican, or heliocentric model contrasted with the geocentric, or Aristotelian model). The 'everyman' character, named Sagredo, begins with no strong opinion either way, while the highly unsubtly named philosopher Simplicio argues for the geocentric model and a third character, Salviati, argues for Copernicus's model.

Over the course of the book, Salviati and Simplicio present the state-of-the-art arguments of the time for or against a heliocentric model. Many of Saliviati's points, perhaps unsurprisingly, refer to Galileo's earlier, very famous observations with the first astronomical telescope (see pages 176–179). They included the first records of spots on the Sun and mountains on the Moon, which were used to demonstrate earthly imperfections in the heavens. Galileo had also previously published observations of the four large (now 'Galilean') moons orbiting Jupiter, and the changing phases and size of the disc of Venus. Such observations are hard to understand in an Earth-centric universe where everything circles Earth: the moons clearly circled Jupiter not Earth, and the phases of Venus demonstrated it could not always remain between Earth and the Sun, it must pass around behind the Sun from our perspective.

These astronomical observations provide the main evidence Galileo presents for the heliocentric model via the mouth of his character Salviati. Contrasting it, Simplicio often brings in scripture and even his scientific arguments (for example referencing a common idea at the time which ridiculed the possibility of Earth moving, because objects fall vertically to the ground) are quickly shot down. In several places he is caught out for errors and often appears as a fool.

Towards the end of the book Sagredo appears convinced by the arguments for the heliocentric model, which despite Galileo's attempts to appear neutral are clearly

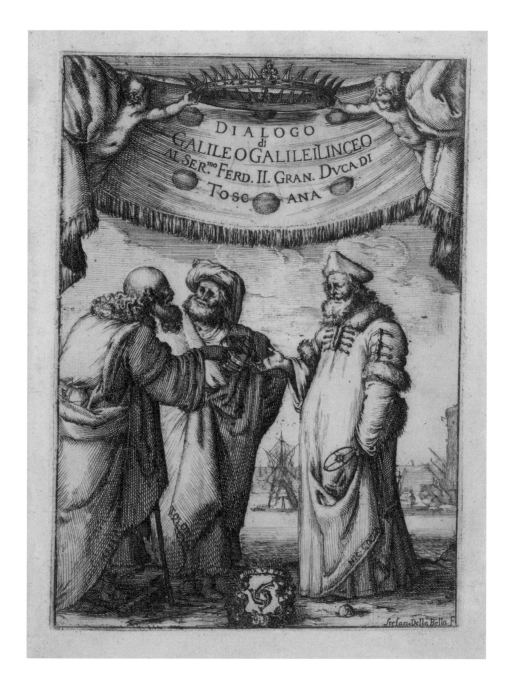

Dialogo sopra i due massimi del mondo

Frontispiece (opposite the title page) of the *Dialogo*, 1632 edition, showing the characters Sagredo, Simplicio and Salviati having their discussion.

presented more strongly that than those supporting the geocentric model. However, as previously agreed with the Pope, Galileo ended the book with an attempt to revert to neutrality. As dictated by the Pope in advance of publication, the closing paragraphs declare that the entire idea of a heliocentric model is probably wrong, and ultimately only God may be able to understand the heavens. At his Inquisition trial, Galileo claimed this was a genuine attempt to avoid arguing for heliocentrism. However, he placed the words in the mouth of Simplicio, who by this point in the book was established as a fool. This angered the Pope and Galileo was declared 'suspect of heresy', placed under house arrest and his book was banned – along with any other book by

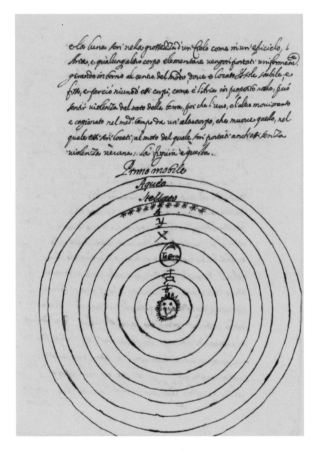

Galileo (including anything he would ever write subsequently!). The future was not kind to this decision. We know today that the Solar System is indeed heliocentric; Galileo was simply speaking the truth. It took about 200 years for the Catholic Church to concede this – it was not until 1835 that they had completely lifted all bans on work in support on heliocentrism (though there were partial lifts in the 18th century), and since the 1990s various popes have issued statements in praise of Galileo's work.

BELOW

Dialogo sopra i due massimi del mondo

This 1632 printed version of the Dialogo was printed by Giovanni Batista Landini in Italy.

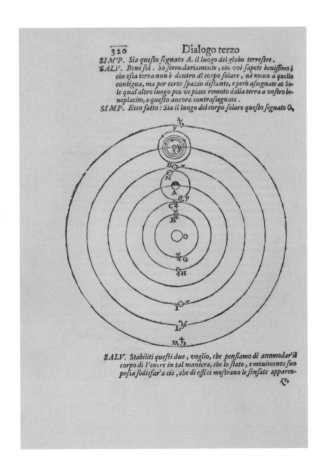

Utriusque cosmi historia

While Galileo was developing the scientific method, finding evidence for a heliocentric universe and getting into trouble with the Catholic Church, other thinkers, such as Robert Fludd (1574–1637), the author of this book, stuck to geocentricism.

Utriusque cosmi historia, published over a decade after Galileo's first telescopic observations and 80 years after Copernicus published his model, presents a solidly geocentric universe. The book, with a full title of *Utriusque cosmi, maioris scilicet et minoris, metaphysica, physica,* a*tque technica histori*a is not strictly astronomical, but it includes some curious and strangely beautiful astronomical diagrams. The title translates to 'The metaphysical, physical and technical history of the two worlds, namely the greater and the lesser', and in the book, Fludd presented his world view and cosmological model and how it could be used to improve the understanding of humanity and medical ailments. Like many of his contemporaries (e.g. Kepler, see pages 168–171) Fludd had scholarly interests which modern eyes would classify as crossing the scientific and occult. Fludd in particular held beliefs about musical harmonies and mathematical ratios revealing special information about nature. He presents a geocentric universe, with circles of earth, water and air included alongside those of the Moon, the Sun, bright planets and the fixed stars, and outside heaven filled with cherubs and angels.

RIGHT, FAR RIGHT, OPPOSITE

Utriusque cosmi historia

The frontispiece revealing the intent of the book, first published 1624, to link the cosmos to understanding of medical ailments. That was a common practice during this period of history.

A diagram presumably showing Fludd's model of the universe, with humans and animals on Earth in the centre surrounded by oceans full of sealife, skies full of birds, then circles with symbols of the planets in the order typical of pre-Copernican geocentric models. Outside the sphere of fixed stars are what appear to be cherubs and angels.

A beautiful engraving showing the make-up of the universe according to Robert Fludd.

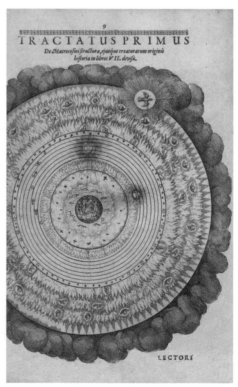

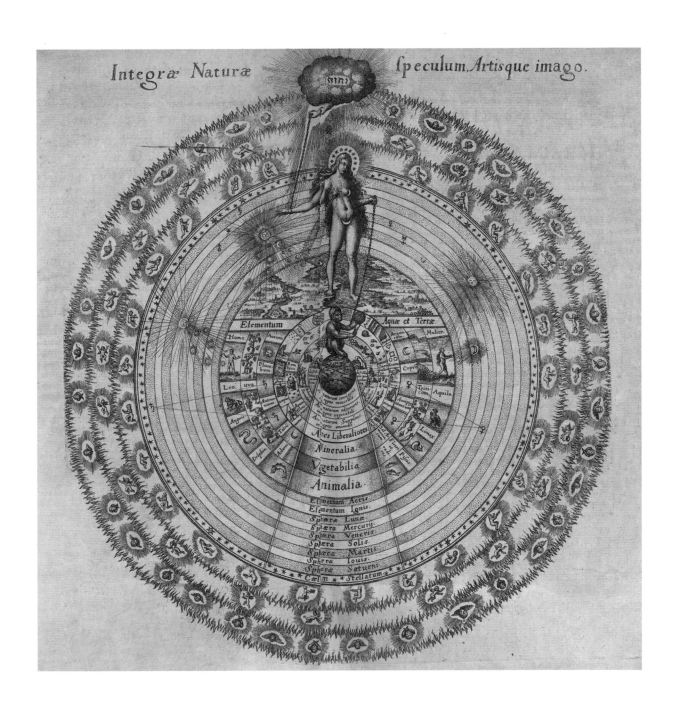

A map of the night sky, which can be set up at an angle presumably as some kind of planisphere (to show where to look for certain constellations at a given time).

The 28-day cycle of phases of the Moon – the spirals indicate the differencing lengths of time the Moon is visible in the night skies (the rest of the time it is in the daytime sky).

Ars magna lucis et umbrae

Ars Magna Lucis et Umbrae, which translates into 'The Great Art of Light and Shadow', is a stunningly illustrated 17th-century exploration of the science of light in a dizzying array of ways humans interact with it. Written by Athanasius Kircher (1602–1680) and first published in Rome in 1646, the book covered material as varied as how fireflies light up at night and how chameleons change colour, and even the nature of colour itself. It discusses the use of lenses and mirrors, both for the projections of images in darkened rooms and for astronomy. We include it here, in part for the section on how humanity could measure the universe using light. There are also several astronomical diagrams showing details of how sunlight, moonlight, and the planets and timekeeping, as well as the typical (for this era) charts for use in medical astrology.

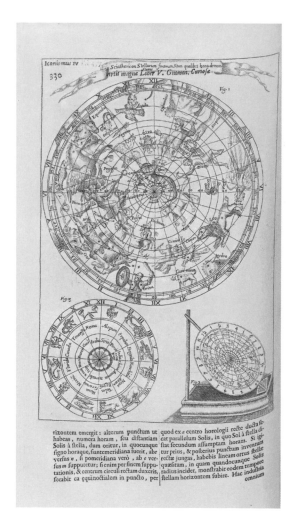

Ars magna lucis et umbrae

A modern image of the
Moon (from NASA). The
bright crater at the lower left
is known as Tycho's crater
(after Tycho Brahe, see pages
172–175), the rays are caused
by ejecta that were thrown out
during its formation.

A detailed sketch of the Moon,
showing many craters, some
with rays extending from them.
This is the inverted (telescopic)
view of the Moon, so the crater
at top is Tycho's crater.

An illustration of the phases of
either Venus or Mercury as they
circle the Sun on orbits interior
to that of Earth.

An early printed image of
Saturn (lower, showing its 'ears')
and Jupiter (upper with two
views of the iconic stripes).

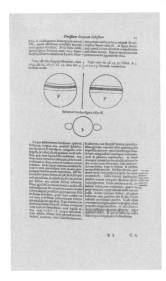

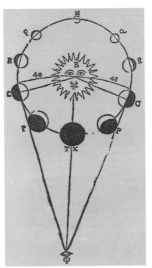

The Ptolemaic geocentric
model for the universe, which
was still popular in 1660 when
Harmonia macrocosmica was
published, despite evidence
Galileo and others had already
published that demonstrated it
could not be true.

Harmonia macrocosmica

Here we note the beautiful maps of the three different – and at the time (roughly 20 years
after Galileo's death) still highly debated – models of the universe. For a larger discussion
of *Harmonia macrocosmica* and its star atlases by Andreas Cellarius and published in
1660, see pages 35–40.

BELOW

Geo-heliocentric model

Tycho Brahe's hybrid geo-heliocentric model as presented in *Harmonia macrocosmica*.

Heliocentric model

The Copernican heliocentric universe as presented in *Harmonia macrocosmica*. This is the most similar to the layout of the Solar System as it is understood today, although planetary orbits are no longer thought to be perfect circles.

Atlas novus coelestis in quo mundus spectabilis

Written by Johann Gabriel Doppelmayr (1677–1750) and first published c.1730 in Nuremberg, Germany, this beautifully illustrated atlas contains 20 richly coloured plates showing aspects of how the universe was understood in the mid-18th century. The most iconic spread from the atlas (see pages 198–199) is a model of the known universe at the time, showing a richly detailed and resolutely heliocentric Solar System. Centred on the Sun, each planet is shown, and those with moons are ringed by the orbits of the moons. The orbits of Mars, Jupiter and Saturn are intricately annotated with additional details. Outside of the Solar System, the rest of the universe is shown by a ring of just the zodiacal constellations around the 'sphere of fixed stars'.

While the circle in the centre shows the main model for the universe, various additional diagrams around the edges of the circle show even more detail of the physical understanding at this time. A whole host of cherubs are found demonstrating how to make use of various astronomical instrumentation including a telescope, sextants and what look like navigational dividers used for measuring distances on charts. At the upper left, a 'to-scale' illustration of the size of the planets relative to the Sun is remarkably similar to today's measurements. Below it we see details of the surfaces of Earth, Mars, Venus and Mercury. Also included are a diagram showing information about a solar eclipse which happened in May 1706, and an illustration of how lunar eclipses occur. At the lower right is a kind of historical reminiscence of different world systems (Ptolemy, Tycho and Copernicus are all shown), meanwhile at upper right, decorating the extended note to the 'Belevole Spectator' (benevolent reader) are numerous heliocentric systems scattered around the clouds. This appears to be a speculation on possible worlds outside of the Solar System; the text describes current understanding of the vast distance to the Sun (describing how a cannonball would even take 25 years to reach it) and that this distance is nothing compared to the vast distance to the stars, which the author says would take a cannonball almost 700,000 years to traverse.

This atlas shows the science of astronomy on the cusp of discovering the enormity of the universe, as it moves beyond the Solar System as 'everything there is' and see space as a potential destination for future travels. It would be another 100 years before the distances to the stars would be confirmed by measurement of parallax (see page 210).

RIGHT

Atlas novus coelestis…

A nice detailed inset showing cherubs working hard to understand the universe using a sextant, maps and measuring instruments.

BELOW

Atlas novus coelestis…

A highly informative page
about the Moon, with detailed
engravings. In the modern
world we would probably call
this an 'infographic'.

ABOVE AND RIGHT
Planets in scale

Scale drawing of the
planets of the Solar System
in Dopplemayer's 18th-
century atlas, compared
to an illustration published by
NASA in the 21st century.

OPPOSITE
Atlas novus coelestis...

A beautifully illustrated
engraving from which includes
highly detailed technical
information about the actual
and apparent orbital motions
of Venus and Mercury. Dates
and angles of observation for
1710 are marked, which show
how the planets appear to have
retrograde motion (from the
perspective of Earth). On each
orbit is also marked perihelion
and aphelion (closest and
most distant point to the Sun
respectively), indicating an
elliptical orbit model is used.
Insets show predictions for the
1710 transit of Mercury and
1761 transit of Venus.

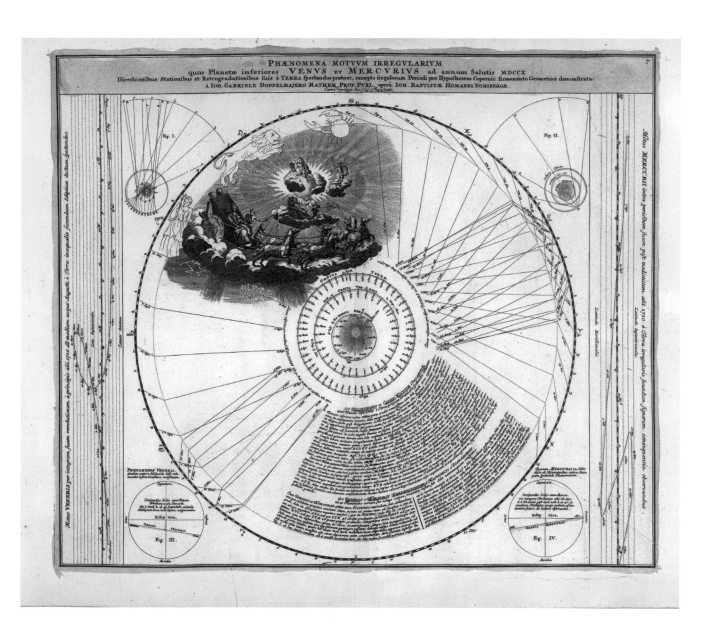

Various planetary and
Solar System 'phenomena',
including the phases of Venus
and Mercury, different views
of the surface of Mars, the
appearance of Saturn through
a full Saturnian year, and
how cloud bands on Jupiter
had been observed to change
over time.

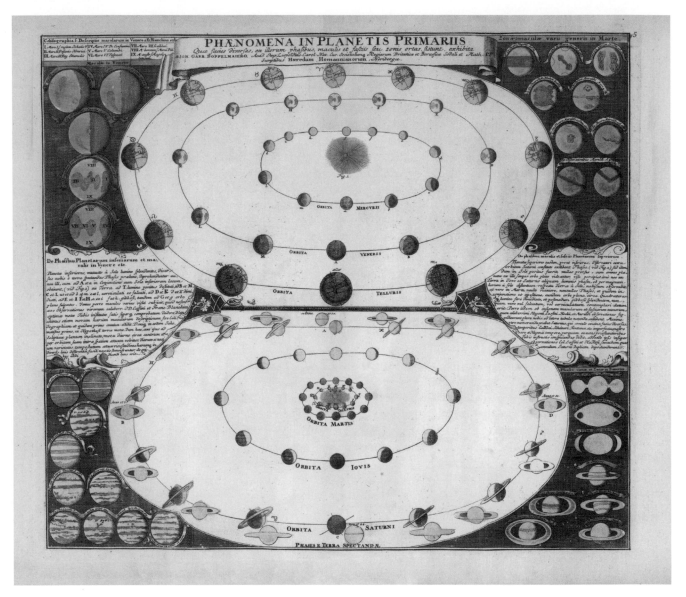

Atlas novus coelestis…

An illustration of the reasons for the length of the day at different times of year and locations on Earth (centre) in a Copernican universe (heliocentric) is shown surrounded by other 'world systems' (ideas about the layout of the Solar System) from both ancient and more modern thinkers.

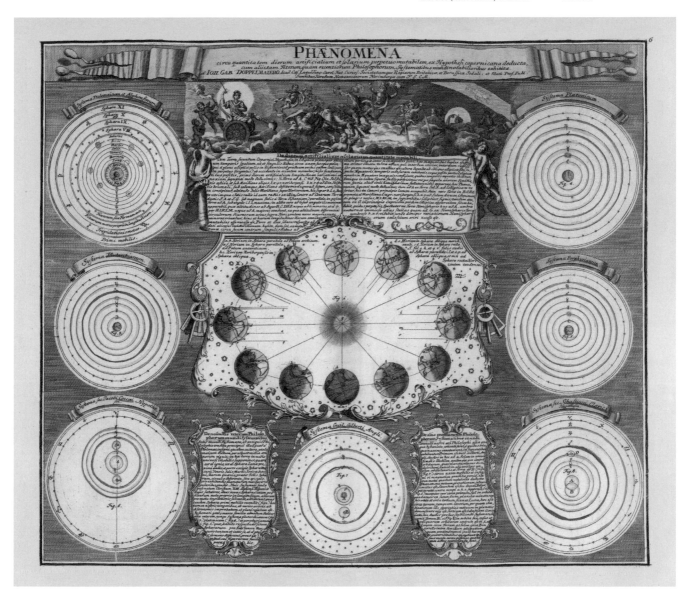

Atlas novus coelestis…

A stunning infographic of the universe, as understood c.1730. In the centre is the heliocentric Solar System surrounded by artwork of the zodiac constellations. The edges include (clockwise from top left): a relative scale of all the planets, drawings of the surface of Earth, Mars, Venus and Mercury, details of a solar eclipse, cherubs illustrating how to use astronomical instrumentation or displaying former Solar System models, how a lunar eclipse happens, then commentary on the possibility of travel to other star systems and the likely scale of the universe outside the Solar System.

BELOW AND OPPOSITE
Principia

Isaac Newton's hand-amended
pages from the first edition
of the *Principia* that he
annotated in the margin and
on interleaved sheets.

Principia naturalis principia mathematica

Isaac Newton (1642–1727) wrote *Philosophiæ naturalis principia mathematica* (The
Mathematical Principles of Natural Philosophy), more commonly known as just *Principia*.
This book has been considered so important to the history of science for so long that it feels
almost impossible to write anything new about it. As early as the mid-18th century it was
being written about as a classic seminal work in science in general, let alone astronomy.
Without the developments laid out in the *Principia*, we simply couldn't do astronomy the
way we do it today. So obviously, this book is a must have in our Astronomers' Library. We
note that three major editions of this work were published across a period of 40 years. The
first – the one we're collecting – came out in 1687. Twenty-six years later, in 1713, Newton
published a version in which he corrected various errors (yes, even Newton made errors!),
the final, definitive version was published in 1726, the year of Newton's death.

Isaac Newton was an English mathematician and physicist most known for the
development of a law of gravity and the laws of motion published in the *Principia*.
Newton also made contributions to the field of optics, including the development of the
first telescopes to use mirrors rather than lenses – something we still call a Newtonian
telescope (see illustration on page 41). Newton is also given joint credit (with Gottfried
Wilhelm Leibniz, 1646–1716) for the development of calculus. He was born into a
minor landed English family and admitted to the University of Cambridge aged 19. Just
six years later, despite the closure of the university during 'The Great Plague' (an episode
of bubonic plague, 1665–1666), he was hired as a fellow, or member, of the faculty. In

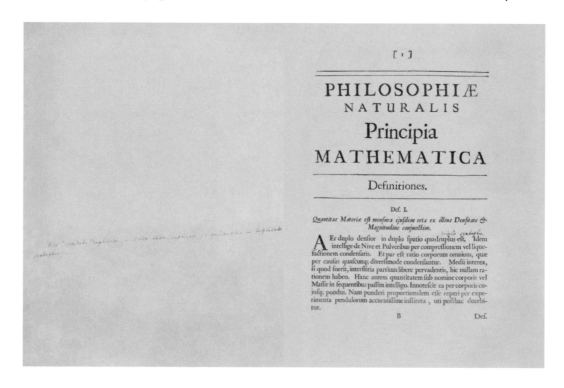

[3]

vim inertiæ. Est autem vis impressa diversarum originum, ut ex ictu, ex pressione, ex vi centripeta.

Def. V.

Vis centripeta est qua corpus versus punctum aliquod tanquam ad centrum trahitur, impellitur, vel utcunq; tendit.

Hujus generis est gravitas, qua corpora tendunt ad centrum Terræ: Vis magnetica, qua ferrum petit centrum Magnetis, et vis illa, quæcunq; sit, qua Planetæ perpetuo retrahuntur a motibus rectilineis, et in lineis curvis revolvi coguntur. Est autem vis centripetæ quantitas trium generum, absoluta, acceleratrix et motrix.

Def. VI.

Vis centripetæ quantitas absoluta est mensura ejusdem major vel minor pro efficacia causæ eam propagantis a centro per regiones in circuitu.

Uti vis Magnetica major in uno magnete, minor in alio.

Def. VII.

Vis centripetæ quantitas acceleratrix est ipsius mensura Velocitati proportionalis, quam dato tempore generat.

Uti Virtus Magnetis ejusdem major in minori Distantia, minor in majori: vel vis gravitas major in Vallibus, minor in cacuminibus præaltorum montium (ut experimento pendulorum constat) atq; adhuc minor(ut posthac patebit) in majoribus distantiis a Terra; in æqualibus autem distantiis eadem undiq; propterea quod corpora omnia cadentia (gravia an levia, magna an parva) sublata Aeris resistentia, æqualiter accelerat.

Def. VIII.

Vis centripetæ quantitas motrix est ipsius mensura proportionalis motui, quem dato tempore generat.

Uti pondus majus in majori corpore, minus in minore; iuq; corpore

B 2

addition to his work in science, Newton was a Member of Parliament, and he led the development of a new currency for England as Master of the Royal Mint. He died in his sleep age 85, still a bachelor.

Newton's publication of the *Principia* is often noted to mark the end of the Copernican revolution, but I think to call this the ending of anything is a bit unfair. *Principia* only ends the Copernican revolution because it marks the start of a new, mathematical scientific revolution. This work was the first large-scale attempt to make use of physical laws and mathematical reasoning to understand the properties and motions of objects in space. It defined the part of modern physics called 'classical mechanics', which deals with the mathematics of the motion of objects. Today, we might call *Principia* a kind of 'astrophysics', meaning an application of physics to astronomical objects, but at the time of the initial publication of the *Principia*, even the definition of 'physics' as a sub-field in science had not yet been made (as you can tell by the use of 'Natural Philosophy' in the full title).

In *Principia*, Newton took Kepler's empirical laws of planetary motion (see pages 168–171) and explained them as natural consequences of a simple law of gravity when it was combined with his three fundamental laws of motion. He described gravity as a force which depends on the mass of objects involved, and as being an inverse-square law. To call it an 'inverse-square law' simply means Newton's gravitational force decreased with the square of the distance between the massive objects – i.e., it gets smaller quickly with distance. Newton's three laws of motion, taught in every introductory physics course today, are that: (1) objects will stay at rest or move at constant speed in straight lines unless forces act on them, (2) any change in momentum (which Newton defined as mass time velocity) is proportional to the force applied, and (3) 'every action has an equal and opposite reaction' (i.e., if you push on something, it pushes back on you). Applied to planets, these rules together mean that to move planets round in their orbits requires a force – gravity, and for the distant planets this force is reduced, so they end up having to change direction more slowly – and, for example, take longer to move around the Sun.

These simple laws, along with new developments in mathematics, allowed Newton to explain a wide variety of observations of the Solar System and to make estimates of the masses of all the planets. Using these methods, he was able to explain not only Kepler's laws of planetary motion but also how the Moon caused tides in the oceans, including why there were spring and neap tides (which involves the relative position of the Moon and the Sun). He could explain why there is a 'precession of the equinoxes' – a slow, 26,000-year cycle of which zodiac constellation the Sun is in at the time of the spring or autumn equinox, which is caused by an effect similar to that of a spinning top wobbling, and even make estimates about the shape of Earth itself – a not-quite-perfect sphere because it gets squashed a bit because it is rotating.

One of the significant developments Newton made, which *Principia* marks part of the start of, is the use of calculus in astronomy (and science in general). This field of mathematics allows the user to more easily keep track of constantly changing variables – for a mathematical astronomer, this could include things like keeping track of the distance from a planet to the Sun and how it impacts on the force of gravity a planet experiences, or to keep track of the constantly changing velocity of a planet as it moves in its orbit. Without calculus, the geometric techniques needed for these calculations are much more complex and unwieldy. It remains somewhat of a debate in the history of science how much calculus Newton used in developing the theories he presented in *Principia*, although he himself is quoted as saying he 'made much use of the method', and it is certainly the case that we teach these ideas with calculus today. However, there is little, if any, calculus printed in the *Principia* itself, which instead makes the arguments geometrically – with numerous geometric diagrams. This was perhaps in part a concession from Newton to make his ideas easier to communicate, since most of his contemporaries at the time would not understand calculus directly.

The most spectacular diagram in the *Principia* is one showing details of the orbit of what at the time was called 'The Great Comet of 1680'. This comet must have been quite a sight in the skies in the autumn of 1680, when it appeared with an extremely long visible tail; reports even say it was visible during the day. After this first appearance, it disappeared for a time, as it rounded the Sun, and reappeared in spring 1681. Initially the spring 1681 comet was taken to be a new object, but the Astronomer Royal at the time, John Flamsteed, (see pages 45–51) proposed that it must actually be a reappearance of the first comet. Newton (eventually) took to this idea, and after also taking Flamsteed's records of the path of the comet across the sky (without permission, as it appears this was done by sneaking into the observatory at night with Edmond Halley), he used the data to provide a spectacular test of his laws of gravity and motion. The diagram shows how the comet follows a parabolic orbital path beautifully.

SECT. II.

De Inventione Virium Centripetarum.

Prop. I. Theorema. I.

Areas quas corpora in gyros acta radiis ad immobile centrum virium ductis describunt, & in planis immobilibus consistere, & esse temporibus proportionales.

Dividatur tempus in partes æquales, & prima temporis parte describat corpus vi insita rectam *A*B. Idem secunda temporis parte,si nil impediret,recta pergeret ad *c*,(per Leg. I) describens lineam B*c* æqualem ipsi *A*B, adeo ut radiis *A*S, B S, *c* S ad centrum actis, confectæ forent æquales areæ *A* SB, B S*c*.[(a)] Verum ubi corpus venit ad B, agat vis centripeta impulsu unico sed magno, faciatq; corpus a recta B*c* deflectere & pergere in recta BC. Ipsi B S parallela agatur *c* C occurrens BC in

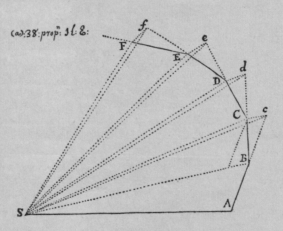

(a):38:prop: 31. 8:

C, & completa secunda temporis parte, corpus (per Legum Corol. I) reperietur in C, in eodem plano cum triangulo A SB. Junge SC, & triangulum SBC, ob parallelas S B, C*c*, æquale erit triangulo SB*c*, atq, adeo etiam triangulo SAB. Simili argumento si vis

[(a) 37: 1:

LEFT

Principia

Example of one of the many pages of the *Principia* showing geometric diagrams as part of the explanation. This page has to do with the effects of centripetal acceleration, felt by any object moving on a curved path.

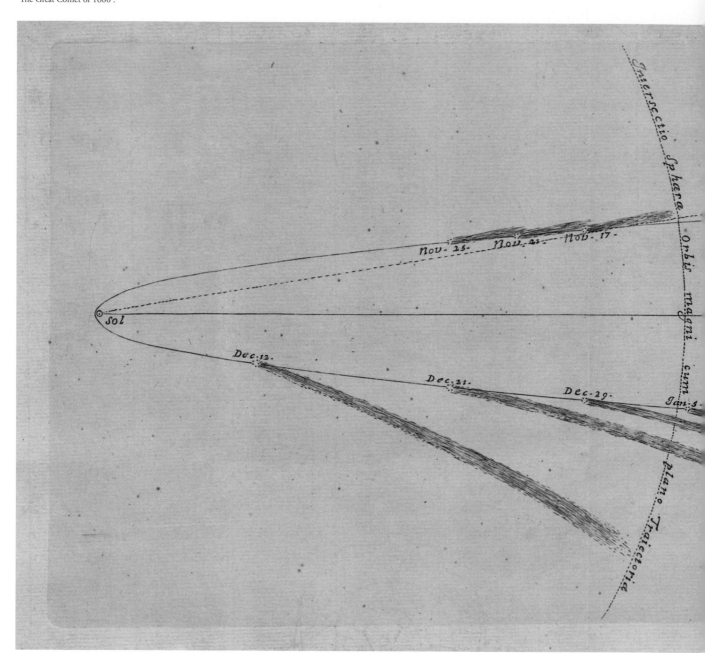

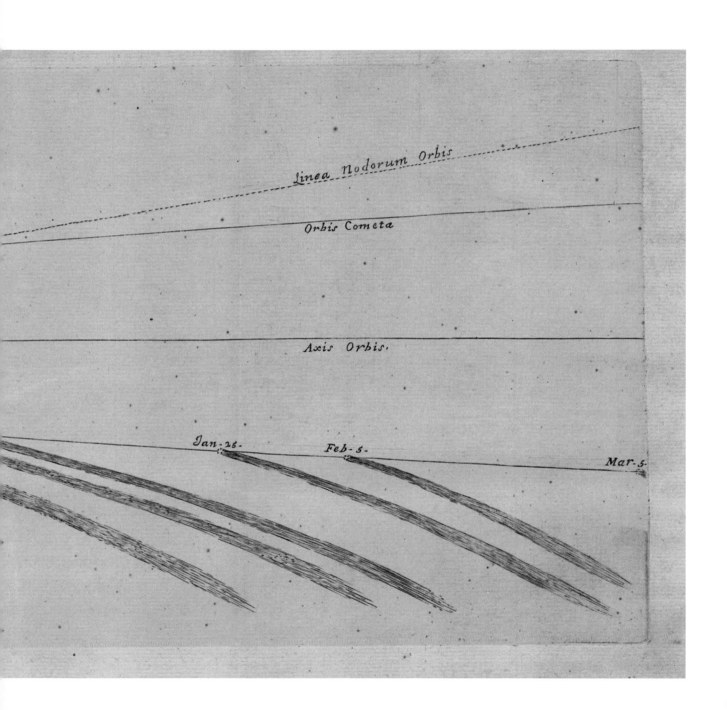

Principes mathématiques de la philosophie naturelle par feue Madame la Marquise du Châtelet

Émilie du Châtelet (1706–1749) was a French mathematician and scientist who was most famous for her translation of, commentary on and additions to Newton's *Principia*, published in 1759. Du Châtelet was born in Paris to a family which was part of the French aristocracy. She was the only girl in a family of six children (some of whom died very young). It is unclear exactly how she was educated as a child, but formal education for girls at this time was rare. However, it is recorded that following her arranged marriage to a marquis when she was just 19 and the birth of three children with him, she resumed her mathematical education at the age of 26. She famously collaborated with the French scientist and philosopher Voltaire (François-Marie Arout, 1694–1778) and while working with him was the first woman to have a paper published by the Paris Academy. She died aged 43, following a difficult childbirth after becoming pregnant during an affair with a French poet.

More than just a simple translation into French, du Châtelet added to Newton's *Principia* the concepts of both kinetic energy (the energy of motion) and that conservation of energy could be used to predict the motions of objects. She also brought more of the calculus of Leibniz (and Newton) to the explanations in her book. Published ten years after her untimely death, this book made significant contributions to the ongoing development of the scientific revolution at the time across Europe.

RIGHT AND FAR RIGHT

Principes mathématiques

Émilie du Châtelet (upper right) depicted as a muse for Voltaire, reflecting the knowledge of the heavens down on him. This is from the frontispiece to Voltaire's commentary on Newton's work, *Éléments de la philosophie de Newton* (1738), co-authored by du Châtelet.

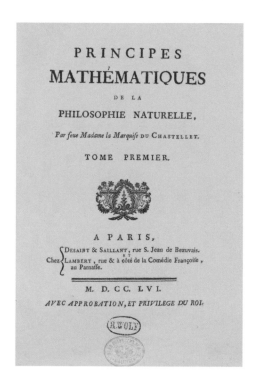

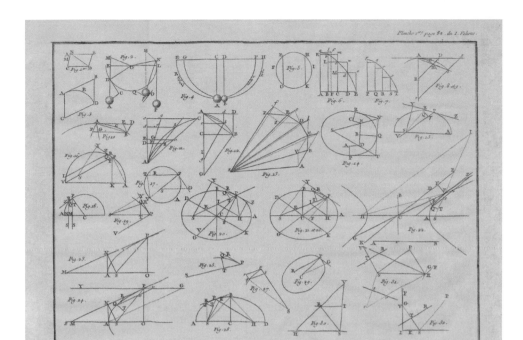

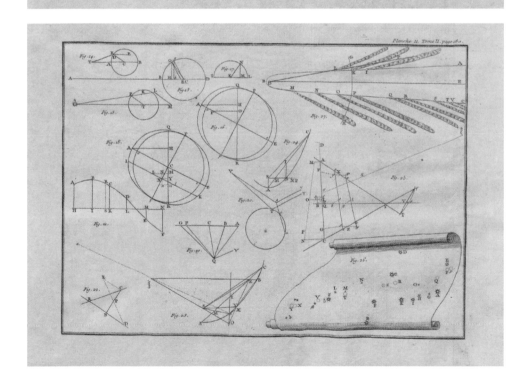

LEFT

Principes mathématiques

In du Châtelet's version, multiples of the geometric figures from the *Principia* are collected together into single plates. Notice the diagrams of pendula (top), and the copy of Newton's explanation of centripetal acceleration (see page 203).

The lower panel includes a small version of Newton's diagram of the comet of 1680, and a chart showing positions of the comet as it rounded the Sun.

Traité d'horlogerie

In the mathematical astronomy era which was launched by Newton's *Principia*, accurate timekeeping would become more and more important. As noted in Chapter 3, timekeeping and astronomy have always been connected, but until this point it was mostly astronomy being used to figure out the time using sundials or by observing the stars. Now astronomers wanted detailed times to record the positions of planets and match them to their detailed mathematical models, and every new 18th-century observatory needed an accurate astronomical clock. *Traité d'horlogerie*, or Treatise of Clockmaking, was published in Paris in 1767 and is an example of this. It provided detailed instructions of how to build accurate clocks, along with tables of the period of oscillations of pendulums of different lengths, presumably to help with calibration.

The book was by Nicole-Reine Lepaute, Jean-André Lepaute and Jérôme Lalande. In the Astronomers' Library we have credited Nicole-Reine Lepaute (1723–1788), a French mathematical astronomer, with co-authorship of *Traité d'horlogerie*, however you may notice the original frontispiece does not. Nicole-Reine worked with her husband, Jean-André Lepaute, a royal clockmaker, and the astronomer Jérôme Lalande on the content of the book, but she was not credited at the time. She is also notable for being partly responsible for the calculations which allowed for a more accurate prediction of the next return of Halley's comet (in spring 1759), which Halley himself could only estimate roughly due to the difficulties of accounting for the gravitational perturbations caused by both Saturn and Jupiter on the motion of the comet.

RIGHT AND OPPOSITE
Traité d'horlogerie

Title page of *Traité d'horlogerie*, where no mention of the (female) co-author is made.

Examples of diagrams of a clock workings.

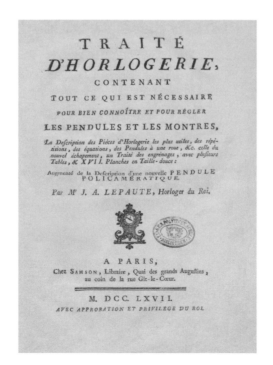

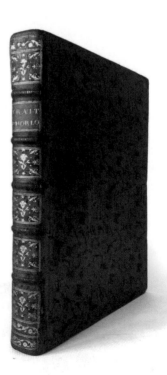

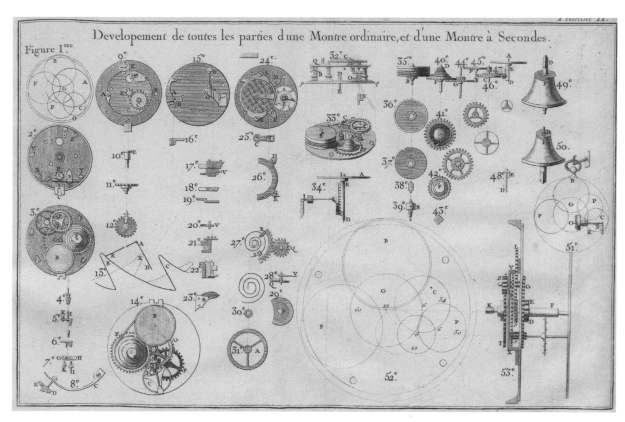

Developement de toutes les parties d'une Montre ordinaire, et d'une Montre à Secondes.

Figure I.ère

Cadrature de Pendule à Repetition.

Planche IV.

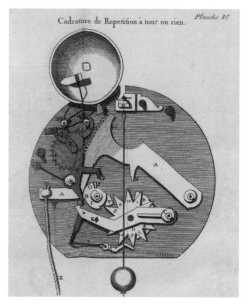

Cadrature de Repetition à tout ou rien.

Planche V.

Transit of Venus

A short pamphlet published in London in 1769, *Transit of Venus* advertised the 3 June 1769 transit – a rare celestial event when Venus passes directly in front of the Sun from the perspective of observers on Earth. Only the 'inferior planets', Mercury and Venus can transit the Sun, and these events follow predictable patterns (see page 98). Transits of Venus happen in a 243-year cycle consisting of a pair of transits separated by eight years, then a gap – the length of which alternates between 121.5 and 105.5 years.

This pamphlet clearly had multiple goals. To start with it aimed to reassure the general public that the transit was not dangerous. There were sections describing both the Sun and Venus, mostly from the perspective of mythology/religion although there are some comments on telescopic observations. But the reason we collect it in this part of our Astronomers' Library is because of how it mentions the utility of this particular transit for understanding the scale of the universe, saying 'Most of the Princes and States of Europe have been at great expense to send learned men into different parts of the world, with proper mathematical instruments to make observations… because it is said, that thereby it will be certainly known how far distant the Sun is from the Earth'.

While mathematical modelling of the Solar System and especially the motions of planets around the Sun was improving immensely by this time, the puzzle of the exact scale of the universe still remained. Distances are always a challenge to astronomers, who observe objects on what still today is referred to as a celestial sphere, looking like they are all in a flat image with no obvious marker for the depth. Relative distances had been reasoned thousands of years earlier (see page 151 on Aristarchus and his work), and orbits and timings can be predicted pretty well from relative measurements, but it is still unsatisfying to not know how big the universe is. To measure a physical scale, the obvious first step was to find an accurate distance from Earth to the Sun, a distance we still call an 'astronomical unit', clearly signifying its important to astronomers. *Atlas novus coelestis* (see page 192) refers to an estimate of 25 years of cannonball flight marking this distance (based on numbers from the Dutch astronomer Christian Huygens, who made it out to be 24,000 Earth radii by comparing angular scales of Venus and Mars).

The idea of using transits of Venus or Mercury to more accurately measure the distance from Earth to the Sun had been first discussed in 1663 by the Scottish mathematician James Gregory (1638–1675). The basic idea is as follows: because the exact timing of the transit depends on the observer's location on Earth, observations of the time of the event from far-flung locations can be used to set up a kind of giant triangle in space. The base of the triangle is known, from the distance between observers, the angle from the time difference these observers recorded for the transit (converted to a 'solar parallax' or an angular difference) and from these two measurements the height of the triangle – the distance from Earth to the Sun – can be calculated.

The 1769 transit was the second that century (1761 was the first). Expeditions had been made in 1761 to make the observations, and there

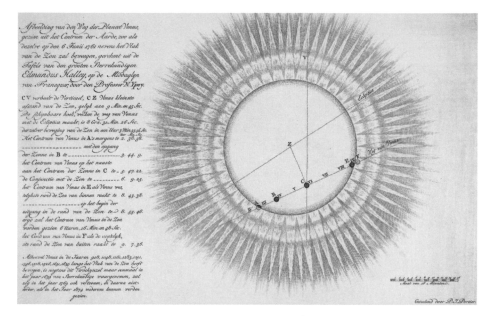

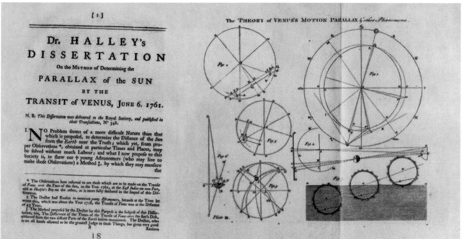

LEFT, BELOW, OPPOSITE

Venus

A drawing of the transit of
Venus from 1761 by Nicholas
Ypey; commentary in Dutch.

Diagram showing the method
of using the transit of Venus
to estimate the distance to the
Sun, by Edmond Halley, 1716.

The *Transit of Venus* flyer, dating
from 1769.

was appetite to do even better in 1769. The 'black-drop' effect (see illustration on page 99) hindered an accurate timing, but even so, detailed mathematical analysis of observations collected from both 1761 and 1769 resulted in a distance measurement of 95 million miles (153 million km), pretty close to the 93 million miles (150 million km) we know it is today. The value was refined further in the 1874/1882 transits, and by the time of the most recent pair of transits in 2004/2012, this method had been superseded by direct radar measurements of distances to Mars and Venus (measuring the light travel time directly for a there-and-back trip of the radio waves). The next pair of Venus transits will be in 2117 and 2125.

Elementos de matematica VII

Another book which is useful to help astronomers grapple with the mathematics they need to develop their understanding of the universe is *Elementos de matematica VII* (or volume seven of Elements of Mathematics) published in 1775 by the Spanish mathematician and mathematical astronomer Benito Bails (1730–1797). Bails wrote a total of 11 volumes for Elements of Mathematics, in which he deals with not only techniques in pure mathematics but aspects of architecture, physics and astronomy as mathematical subjects. In the foreword to Volume VII, which is *Elementos de astronomia*, Bails comments that in his view the application of mathematics to astronomy has *el del mayor elevacion* (the highest elevation).

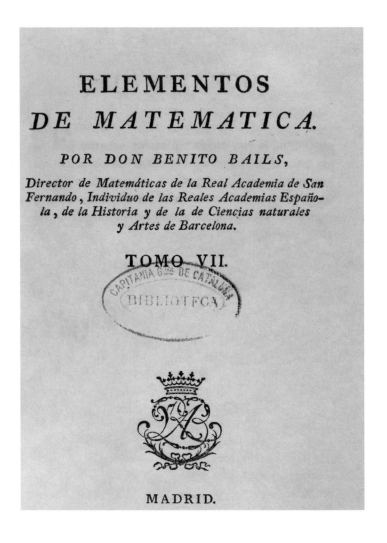

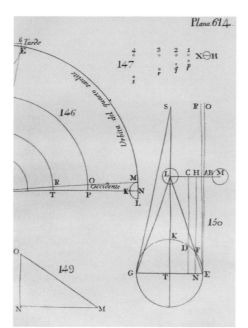

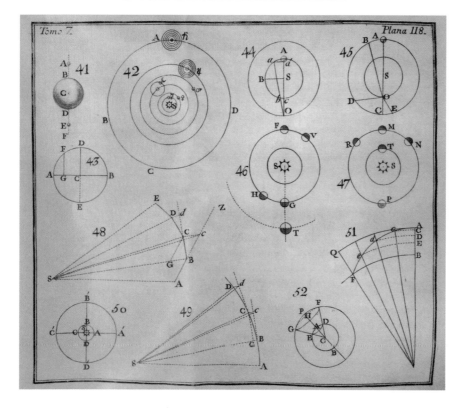

LEFT

Elementos de matematica

A selection of diagrams from the book, showing various mathematical and astronomical concepts. You can find models and diagrams for demonstrating celestial co-ordinate systems, and illustrations of the phases of Venus, among other diagrams.

Abrégé de navigation historique théorique et pratique avec tables horaires

Abrégé de navigation historique théorique et pratique avec tables horaires (A Historical, Theoretical and Practical Summary of Navigation with Tables of Timings), is collected here as an example of a book describing techniques for astronomical navigation for the 18th-century seafarer. The author is listed as the astronomer Jérôme Lalande and the book was first published in 1792 in Paris.

Developing an understanding of the universe we live in isn't just interesting in itself, but it can also provide economic benefit, for example, via increasingly accurate methods for navigation, especially navigation at sea. In the era of global navigation satellite signals, it is hard to remember just how hard this was for much of human history.

While for any decent astronomical navigator, figuring latitude based on observations of the height of the North Star (the star which happens to currently be close to the location of the north celestial pole) or the south celestial pole, is very easy,[2] longitude (or position east–west) around Earth is much trickier.

Cultures across the world practised sophisticated methods of celestial navigation. As early as 900 BCE, Polynesian navigation – partly based on celestial navigation – was so advanced that a distinctive culture arose crossing more than 3,700 miles (6,000 km) across the immensity of the Pacific Ocean. There is evidence that Ancient Europeans also used celestial navigation across the much smaller Mediterranean Sea, including records in some of the Classical texts (e.g. Homer's *Odyssey* mentions celestial navigation). In the Islamic golden age, the development of magnetic compasses and astrolabes, along with early precursors of the sextant, aided navigators.

Around 100 years before this book was published, astronomical observatories had been built in both Paris and London (the Paris Observatory and the Royal Observatory at Greenwich were founded in 1667 and 1675, respectively), in large part funded by the royals in both England and France (at that time in the midst of one of the many Anglo-French wars) because they wanted to improve methods and data for nautical navigation.

The calculations were complex. Marie-Jeanne de Lalande (1768–1832) was a French mathematical astronomer, and her astronomical calculations were used in the writing of *Abrégé de navigation historique théorique et pratique*. She contributed to many other works on astronomy, but at the time she was never listed as a co-author. We know of her contributions only due to extensive mentions in prefaces by Joseph Jérôme Lefrançois de Lalande (1732–1807), a French astronomer and prolific author of astronomical books, who was a vocal advocate for women in astronomy at the time. Some records have

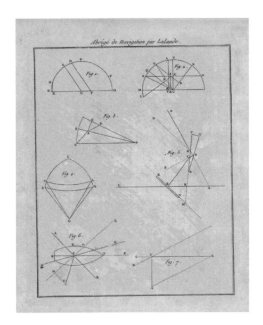

suggested Marie-Jeanne de Lalande was his illegitimate daughter, however recent research suggests this was unlikely to be true, and the link was simply via Marie-Jeanne marrying the son of one of Jérôme Lalande's cousins.

Over the next decades the star charts used in nautical navigation were much improved, however reckoning longitude remained fraught with error. Because of the regular rotation of Earth, which makes sunrise progressively later as you move west, longitude can be estimated by the change in sunrise/sunset times relative to a clock tied to any fixed point. However, clocks which could run accurately in the wet and wobbly conditions found aboard a ship were not possible, at least not until John Harrison (1693–1776) came up with his series of 'marine chronometers' – clocks based on spring mechanisms. This innovation revolutionized both navigation and astronomy, making observatories into centres of time; the ball-drop in New York City's Times Square on New Year's remains as a relic of daily ball-drops from observatories in coastal locations, which navigators on board ship could use to set their clocks before leaving harbour. However, it took a long time for clock-based navigation to be fully adopted, and complex tables of lunar positions (the Moon being close enough that precise position measurements can reveal longitude) were still included in nautical navigation books until at least the end of the 19th century.

---

[2] The angular height the north (or south) celestial pole makes above the horizon gives the observers latitude north or south of the equator – for example, if you stand on Earth's North Pole, the north celestial pole will be directly overheard, while if you stand on the equator, both celestial poles touch the horizon.

Mechanism of the Heavens

While Newton had laid out the basics of the theory to be used to explain the motions of planets in the Solar System in his *Principia* in 1687 (see pages 200–205), much work was still needed to fill in the details of the model.

Pierre-Simon Laplace (1749–1827) took Newton's basic model and refined it (in the view of many completed it) in his *Méchanique celeste*, published in five volumes in French across 1798–1825. In this work, Laplace summarized decades of his own careful analytic work building on Newton's theories. One huge innovation was the addition of a delay in the effect of gravity – today we know that changes in gravity travel at the speed of light – but Newton initially assumed gravity would act instantaneously, which has the effect of making the Solar System appear unstable. Laplace demonstrated by adding a delay, the timescale on which the Solar System is stable is much (much) longer. Laplace also developed a dynamic theory of tides, which better explained the impact of tidal forces in the Earth–Moon system, and developed many mathematical techniques to better allow for the application of calculus to the motions of objects in space.

In her *Mechanism of the Heavens* (published in 1831), the Scottish astronomer Mary Somerville (1780–1872) took Laplace's work and translated it into English. She did not make the first English translation, but she translated the book in a way which made the explanations much clearer, rephrasing many of the mathematical sections to improve the

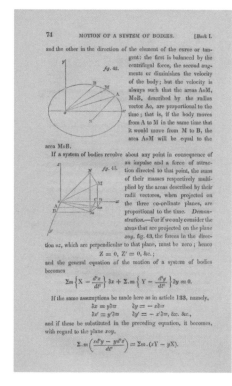

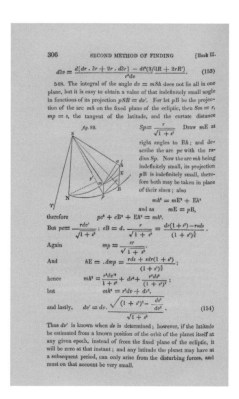

explanations, and her text proved immensely popular. It became the primary calculus textbook in English universities for over 50 years, and Somerville herself became famous as a mathematical astronomer and prolific author of books on a dizzying array of scientific topics. This is despite the fact that Somerville was largely uneducated as a child (typical for girls at that time), and it is said she first discovered mathematics in a women's fashion magazine at a tea party. *Mechanism of the Heavens* had an extensive preface with some 'preliminary observations on the nature of the subject'. This was later expanded into Somerville's *The Connection of the Physical Sciences*, which was published as a stand-alone short book, considered by many to be the first ever popular science book.

In addition to her clear explanations of celestial calculus, in *Mechanism of the Heavens* Somerville defined the set of subjects we now call physics, and a review of it is the first recorded use of the word 'scientist' (since Somerville could not be described as a 'man of science'). She went on to write a number of other popular science books, and become a vocal advocate for the education of women in the United Kingdom.

The methods laid out in *Mechanism of the Heavens* could be used not only to predict the motions of the planets to great precision, but also to notice that the motions of some planets just didn't quite fit the clockwork pattern. When Mary Somerville was just one year old in 1781, the known 'universe' had grown with the discovery of a new major planet, Uranus. Uranus is so faint and moves so slowly through the background of fixed

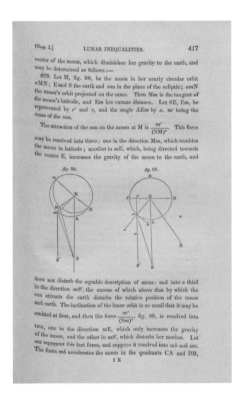

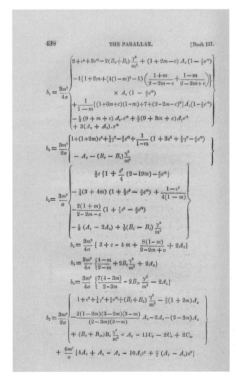

stars, that it had not been possible to notice it was a planet until the era of frequent telescopic observations. By the time Somerville was working on *Mechanism of the Heavens*, there were almost 50 years of observations of Uranus on record. Somerville noted the odd behaviour in its orbital motion and suggested that it could be caused by a missing planet pulling on it with unknown gravity. The English astronomer who calculated the location of this missing planet (John Couch Adams, 1819–1892) entered Cambridge just a few years after the publication of *Mechanism of the Heavens* so would have studied calculus using it as his textbook. He has been quoted as saying he was inspired to do the laborious calculations by her words. A French astronomer, Urbain Le Verrier (1811–1877), had simultaneously made the same calculation, and as the planet Neptune was discovered close to the positions they separately calculated, they are usually given joint credit.

The discovery of Neptune in 1846 could be used to mark the true beginning of mathematical models being used to discover new information about the universe. It is the first example of a planet discovered solely by the impact of its gravity – today over 5,000 extrasolar planets (planets around stars other than our Sun) have been discovered using similar methods. Most exoplanets cannot be seen directly, but they are known to be there by the gravitational effect they have on their star.

Meanwhile, following the development of these techniques, it was recognized that the universe was expanding – literally and figuratively. In 1832, just a year after the publication

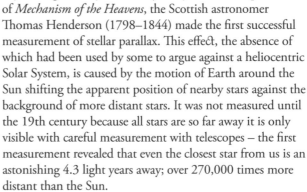

of *Mechanism of the Heavens*, the Scottish astronomer Thomas Henderson (1798–1844) made the first successful measurement of stellar parallax. This effect, the absence of which had been used by some to argue against a heliocentric Solar System, is caused by the motion of Earth around the Sun shifting the apparent position of nearby stars against the background of more distant stars. It was not measured until the 19th century because all stars are so far away it is only visible with careful measurement with telescopes – the first measurement revealed that even the closest star from us is an astonishing 4.3 light years away; over 270,000 times more distant than the Sun.

The subsequent two centuries have revealed a universe full not only of stars, but galaxies of stars (immense collections of stars) which stretch to unimaginable distances, and which reveal the expansion of the entire universe itself. As the time of writing, the record for the most distant known galaxy (from observations with NASA's latest space telescope) is over 30 billion light years. The light from that galaxy began its journey to Earth when the universe was just a fraction of its current age. We can only wonder what discoveries are yet to come about the amazing universe we live in.

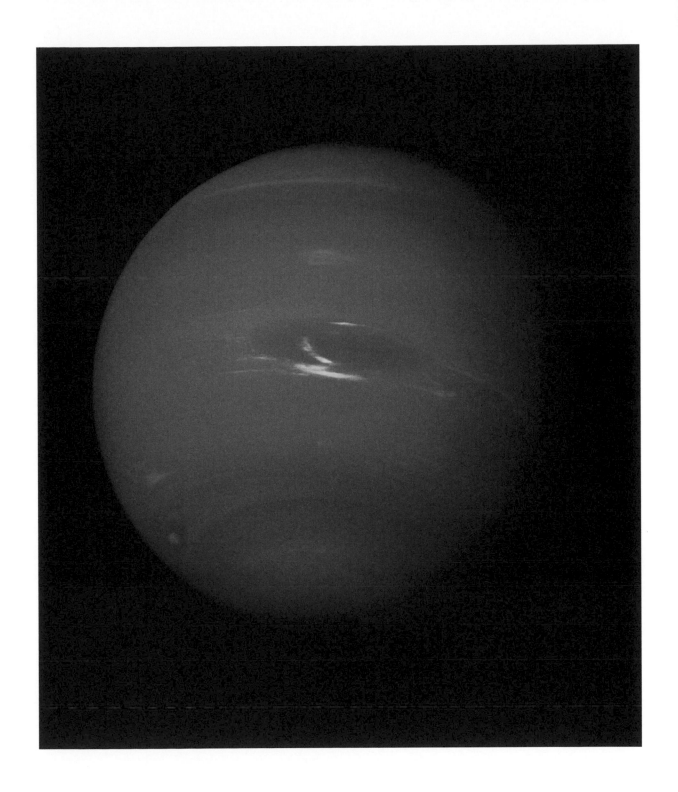

ASTRONOMY FOR EVERYONE

Astronomy is often considered to be a gateway subject – an inspirational, awe-inspiring and often beautiful topic which provides a way to bring people in and develop an interest in science more generally. There's just something about 'outer space' that is endlessly fascinating to children and adults alike. What's more, astronomy is accessible. Anyone can step outside on a clear night, gaze up at the skies above and wonder.[1]

Perhaps it is no surprise then, that for hundreds of years, the majority of books on astronomy have been published with the intent to educate and inspire non-astronomers about the heavens above. In this section of our Astronomers' Library, we'll collect just a few of the many thousands of these books from medieval times up until many of the prolific 'Victorian' authors at the beginning of the 20th century, saving more recent books for Chapter 6: Modern Astronomy. These books are often beautiful, and it is fascinating to look back on them to see how they capture epochs in humanity's developing an understanding of astronomy.

There's no way for this chapter to be complete, no possible way to cover all books ever published to educate about astronomy, so we unapologetically cover a set which have enjoyable illustrations or are particularly notable in other ways. Frankly, most of these books are just fun to look at. And even though in this section, our library is highly dominated by the Anglosphere, they still astonish me in how inclusive they are in bringing astronomy to everyone. In a subject often understood to have been dominated by white men, many of these books since the middle of the 19th century were written by, or for, women.

Our collection of educational astronomy books will include Sacrobosco's *De sphaera mundi* (On the Sphere of the World), written in the 12th century, and arguably the most

---

[1] At least unless light pollution has wiped out the night skies completely.

PROMINENT ASTRONOMERS OF FORMER TIMES.

227.

Agnes Giberne's *The Story of the Sun, Moon and Stars* pictured some of the prominent astronomical figures from throughout the ages. On a later page, more modern scholars were pictured.

The romance of the stars

A 19th-century view of classical astronomy sees Hipparchus watching the skies with a cross-staff – a pre-telescopic tool used in astronomy.

HIPPARCHUS IN THE OBSERVATORY OF ALEXANDRIA.

successful book in astronomy ever. It also includes books known to have inspired significant astronomers, for example, we collect the 1764 book *Astronomy Explained upon Sir Isaac Newton's Principles* by James Ferguson, which is said to have introduced astronomy to William Herschel (who then went on to discoverer Uranus and made the first map of our Galaxy – see pages 144–145 – among other things). Many of the books covered in other sections could also belong here. Mary Somerville's *Mechanism of the Heavens* (see pages 216–218) was a textbook for generations of Cambridge University students in mathematical astronomy and may have inspired one of the discovers of Neptune to make his calculations. Other examples include the short text, *Transit of Venus by* Chintamanni Ragoonatha Chary (see page 99) which was explicitly written (in seven different Indian

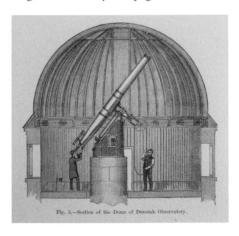

Fig. 3.—Section of the Dome of Dunsink Observatory.

languages as well as English) to educate people in India about the upcoming 19th century transits of Venus, or *Kashf al-Ghummah fi Nafa al-Ummah*, (*The Important Stars Among the Multitude of the Heavens*; see pages 112–113) which we collected as a record of knowledge of astronomy in the medieval Islamic world, but it was at the time written as an educational text. If you think about it, most astronomical books were published to educate someone about astronomy (even if in some cases it was fellow astronomers), so they could all belong here.

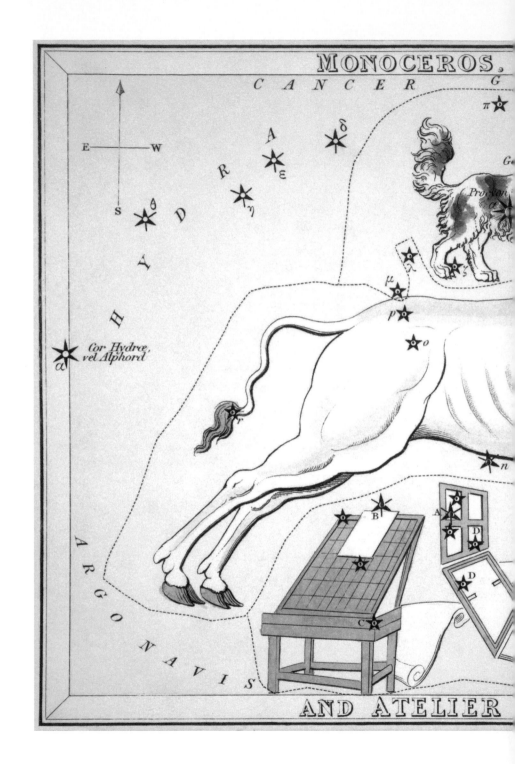

A set of printed cards illustrating the constellations that were very popular upon release in 1825; they remained in print for many decades. In addition to the charming illustrations, each card had tiny holes punched in it at the position of the stars, and so when held up to a light they simulated the night skies. This card shows Monoceros (the Unicorn), a faint constellation near Orion. Also pictured is Canis Minor (the little dog), and 'Atelier Typographique' (printing office), a now-obsolete constellation which was suggested in the late 18th century to honor the printing press of Gutenberg, but was not adopted as an official constellation.

De sphaera mundi

Arguably the most successful book in astronomy ever, this charming medieval textbook on astronomy was used in universities across Europe for hundreds of years. So many copies of *De sphaera mundi* (On the Sphere of the World) were made, that despite it being almost 800 years since it was written, and over 500 years since the first print edition, hundreds remain in good condition to the present day, and perhaps as many as 200 different editions exist, with many subtle (or not so subtle) variations in the illustrations.

Much of the biographical information available on the author, Sacrobosco, is unclear and the years of his birth and death are estimated. Even his name appears murky. Sometimes he is Ioannes de Sacro Bosco, and sometimes John of Holywood/ Holybush (*sacro bosco* translates to 'holy bush'). He may or may not have been born in Britain – several locations (in Northern England, Scotland and Ireland) have been suggested. He may or may not have been educated at Oxford University, which was founded in 1096. What is clear is that at some point he arrived at the University of Paris, founded in 1150, and there he taught astronomy/astrology and other mathematical topics. He is buried in a monastery in Paris.

Sacrobosco based *De sphaera mundi* on the ideas of Ptolemy's *The Almagest* (see pages 152–157) as well as various Islamic commentaries on that same book. He presented a universe with two realms: Earth, the 'elemental realm', consisting of earth, water, air and fire and the Moon, which marked the start of the immutable ethereal realm above, containing the five bright planets and the Sun – beyond that was the sphere of fixed stars.

While the 'sphere' that Sacrobosco referred to in the title was that of the heavens, the book still clearly demonstrates that Earth was known to be a sphere at this time, and shows the interested reader how to reason that for themselves (see illustration on page 142). *De sphaera mundi* presented a model of the universe as a kind of clockwork machine, demonstrating how eclipses occur and noting how they disrupted the normal order of the universe.

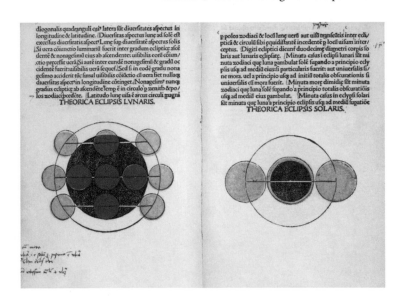

De sphaera mundi presents a solidly geocentric view of the universe. It was likely first published around 1230, and even 200 years later, at the time of the first actual printing in 1472, it would be another 70 years until Copernicus would publish details of his heliocentric model (see pages 164–167). In any case, *De sphaera mundi* remained a popular text well after the Copernican revolution was under way and much of its content was outdated.

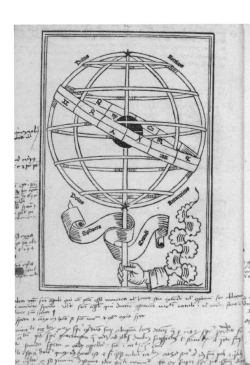

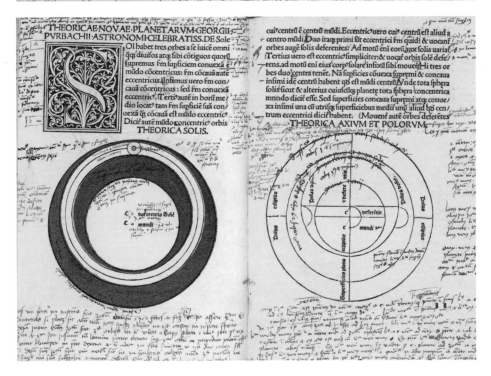

De sphaera mundi

A much annotated edition of *De sphaera mundi*, first published 1472 in Ferrara, Italy.

An illustrator's idea of Sacrobosco holding his sphere of the world made in 1584 – over 300 years after his death.

De sphaera mundi presented a model of the universe as some kind of clockwork machine, this diagram demonstrating how eclipses occur.

Ringed by illustrations of the constellations of the zodiac, mythological figures representing the Sun and planets march across the skies.

An illustration of a geocentric view of the universe, with Earth surrounded by the Moon, and what looks like a comet, then the Sun and the fixed stars. Outside of the fixed stars is the realm of angels.

This panel of the illustration of the Creation shows God making the skies and the stars, and the use of a navigational compass for measurements during this work.

Le livre de la proprieté des choses

A popular medieval European encyclopedia, *Le livre de la proprieté des choses* (Book of the Properties of Things), was initially written in Latin around 1240; this French translation from about 200 years later is spectacularly illustrated; it is a classic example of an 'illuminated manuscript' of the era, and a copy of it sold for over one million dollars in 2010. The entries across 19 separate 'books' provide both a record of knowledge of the time, from the mundane (e.g. daily life) to the divine (e.g. God, angels and the soul). The astronomical-related entries include a book on 'earth and the heavenly bodies' and one on 'time and motion'. As expected from the date of its original publication, this presents a solidly geocentric world view.

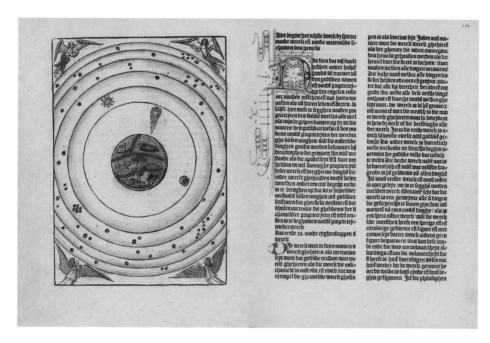

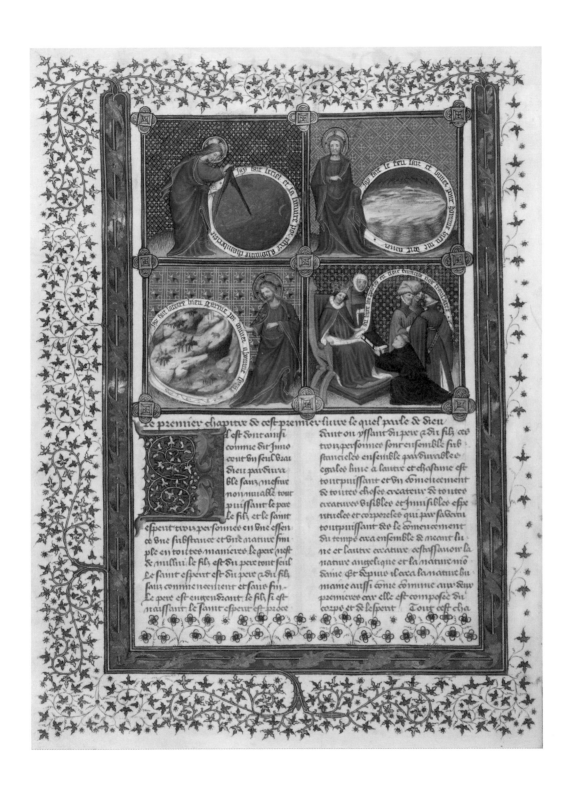

Astronomy Explained Upon Sir Isaac Newton's Principles

Isaac Newton presented his revolutionary ideas in *Principia* in 1687 (see pages 200–205), but that work, like many other technical texts throughout history, needed some expert translation to be more accessible to a general audience. *Astronomy Explained Upon Sir Isaac Newton's Principles*, by the self-taught Scottish astronomer James Ferguson, is a delightfully illustrated example of this genre. The full title is *Astronomy Explained Upon Sir Isaac Newton's Principles, and Made Easy for Those Who Have Not Studied Mathematics*. It was published in London in 1764 and it is said that William Herschel (1738–1822; see page 142) studied from this text before going on to make many contributions to astronomy with his sister Caroline Herschel (1750–1848) – also

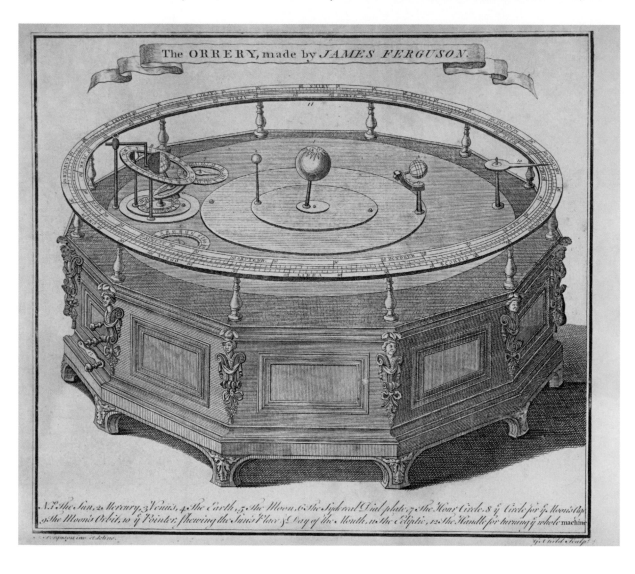

a notable astronomer in her own right. Together the brother-sister team built and polished multiple telescopes, using them to discover many comets, the planet Uranus and to make the first three-dimensional map of our own Galaxy of stars.

The author of this notable book, James Ferguson (1710–1776), was born of humble origins as the son of a farmer in Scotland. In his early life he was a sheep farmer himself and also worked as a servant for a time. He was entirely self-taught in both astronomy and mechanics. After showing a clock he had made to a 'gentleman', he became employed in that house, eventually developing a skill in painting miniatures, which allowed him to move to Edinburgh and make enough money to support his ongoing scientific studies. In his thirties he moved to London and became famous for his complex mechanical models of astronomical systems, eventually lecturing on astronomy.

Ferguson's book contains chapters on state-of-the art astronomy in the 18th century, from a description of the Solar System, the physical causes of the motions of the planets, and the evidence that the Copernican model is correct and Ptolemy's wrong, to the principles of light and the basics of how calculate the timing of eclipses and other astronomical phenomena. It also included very recent updates to the scale of the Solar System, based on observations of the transit of Venus just a few years earlier in 1761 (see pages 210–211).

OPPOSITE AND BELOW
Astronomy Explained…

The book was first published in London in 1764. Author James Ferguson was particularly known for his mechanical astronomical models. His book includes instruction in how to build complex mechanical models like the Ferguson's orrery shown here.

A map of our Solar System and a diagram showing relative planet sizes.

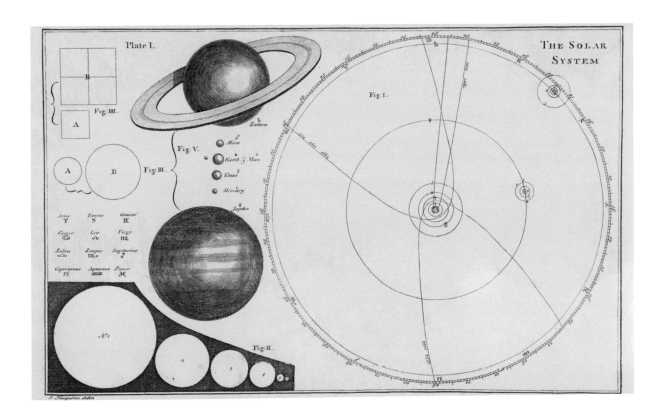

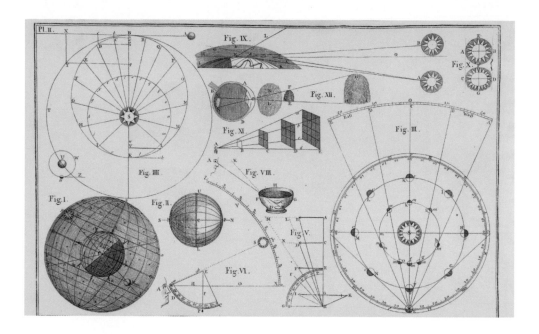

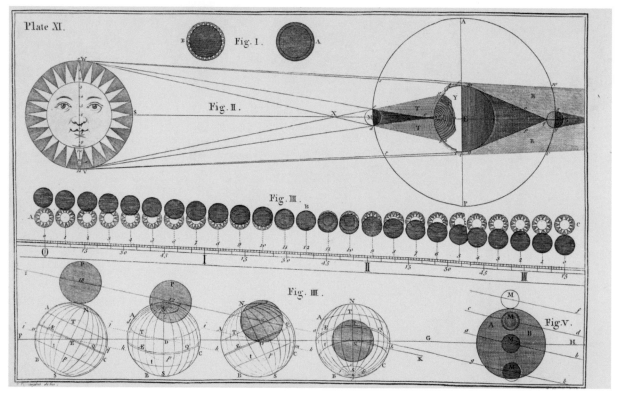

Caroline Herschel.

In this illustration, Fig III demonstrates the phases of the inner planets (Venus and Mercury), among other things.

A detailed diagram showing the progression of a solar eclipse.

An illustration of William and Caroline Herschel from 1896 (some years after they lived). William polishes a telescope mirror, while Caroline assists by providing the powder from what looks like a teacup It is said that William first studied astronomy from Ferguson's book.

A Compendius System of Astronomy

Initially published as a series of small pamphlets, *A Compendius System of Astronomy* is a textbook on astronomy for young women by the schoolmistress Margaret Bryn (c.1760–1815). This was Bryn's first textbook, published in 1797, and she went on to publish two additional texts on 'natural philosophy' (which we'd now refer to as physics) and a book for younger children on astronomy and geography. The compendium includes an image of Margaret Bryn with two of her children surrounded by astronomical instrumentation, and it was dedicated to the 'young ladies' who were her pupils.

RIGHT, OPPOSITE

A Compendius System…

(Clockwise from top left): Margaret and her daughters from the original 1797 edition of the book.

Diagram showing the scale of the planets and their orbits, as well as the apparent size of the Sun as seen from various planets. In this diagram, the planet we now know as Uranus, which had been discovered less than 20 years before the publication of this book, is shown with the initial name given to it by William Herschel: 'Georgium Sidus'.

Diagram showing a model which can be used to understand the motions of the skies around Earth, and other diagrams related to the daily or annual changes of the skies from our perspective.

(Clockwise from top left): A selection of diagrams from the book: a mechanical orrery; map of the Moon and details of its phases; celestial globe; a telescope.

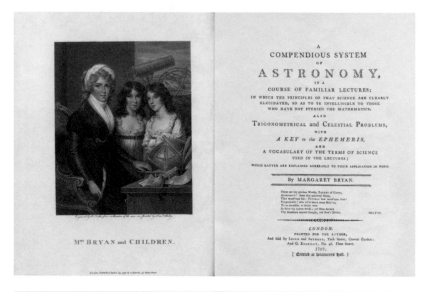

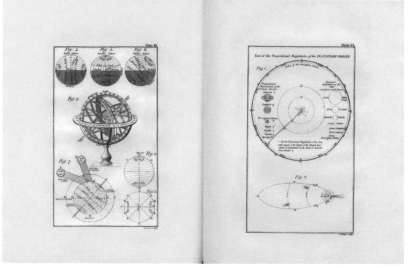

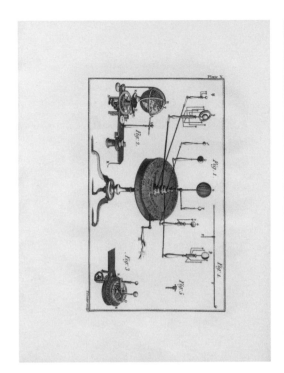

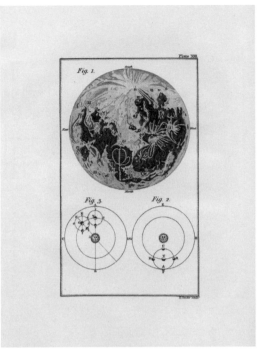

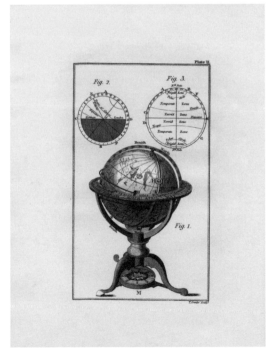

Urania's Mirror; or, A View of the Heavens

This book, published in London in 1825, was written by Jehoshaphat Aspin (1788–1857) as a companion text to a charming set of astronomical star chart cards that were more commonly known as *Urania's Mirror*. Each of these cards has colourful illustrations of the mythological figures that represented the constellations in 32 different patches of the sky,[2] and each card was punched with holes at the locations of the stars, which varied in size with the brightness of the stars, so when held up to the light, they looked like the night sky itself.

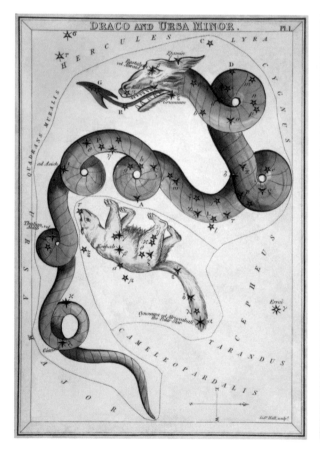

---

[2] Which are suspiciously similar to the artwork in *Celestial Atlas*, an 1822 star atlas by the Scottish schoolmaster Alexander Jamieson – so much so that they are often noted as being direct copies from that work.

ABOVE, LEFT, OPPOSITE

Urania's Mirror

The card for the constellation of Ursa Major.

A rather charming newspaper advertisement for the cards from first release.

Cards showing the illustrations for Draco and Ursa Minor (left) and 'Gloria Frederici' (a now discarded constellation created in 1787 to honour Frederick the Great of Prussia), Andromeda and Triangulum (right).

Illustration from the book's
charming title page.

Part of a lesson intended to
give a schoolteacher a set of
questions for their students.

An illustration of the relative
size of the planets.

Discussion of Mars and other
'planets' from a later chapter.
From their discovery in the
early 1800s, the objects Vesto,
Juno, Pallas and Ceres were
usually considered planets,
although the term 'asteroid'
had been invented to describe
them, it was not yet in common
use until around 1860. Today
Vesto, Juno and Pallas are
known as asteroids, although
since the 2006 reclassification
of Pluto, Ceres is now also
considered a dwarf-planet.

First Book in Astronomy

This short book, illustrated by numerous steel plate engravings, was written as a book for education about astronomy in 'the Common Schools' and published in 1831. As such it is divided into a series of short lessons, clearly intended to help schoolteachers plan their teachings on astronomy. The book ends with a dictionary of astronomy.

The author, Reverend John Lauris Blake (1788–1857), wrote numerous educational texts across a wide range of topics, from geography to art, farming, Christianity and the history of Native Americans, so presumably he was not an expert in astronomy. Nevertheless, this is a lovely example of a children's textbook on astronomy from this era.

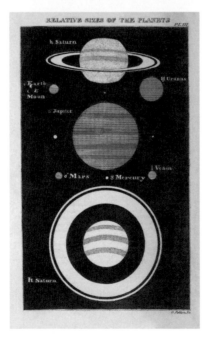

28   FIRST BOOK IN ASTRONOMY.

Mars revolves round the sun in the period of 687 days, or a little less than two of our years. Its mean distance from the sun is 144,000,000 of miles; moving in its orbit at the rate of 55,000 miles an hour.

This planet sometimes rises before the sun, and becomes a morning star. Sometimes it sets after the sun, and then becomes an evening star. It is also seen, at other times, in the same part of the heavens with the sun, and beyond that luminary, being in conjunction.

The inhabitants of Mars have three interior planets, Mercury, Venus, and the Earth. Each of these will sometimes be a morning star, and sometimes an evening star; although the earth will be the brightest and most luminous.

Summer and winter at Mars, are each about twice as long as they are on the earth; and the proportion of light and heat received from the sun is less than one half of what is enjoyed at the earth.

VESTA, JUNO, PALLAS, AND CERES.

LESSON XIV.

These four planets revolve in orbits between those of Mars and Jupiter. They have been recently discovered, and are so very small compared with the other primary planets, that they are frequently called asteroids; or bodies having the appearance of stars.

Ceres was discovered in the year 1801; Pallas, in 1802; Juno, in 1804; and Vesta, in 1807. Ceres was discovered by Piazzi, an Italian; Juno, by Mr. Harding, a German; and Pallas and Vesta, by Dr. Olbers, of Bremen.

It has been supposed that these bodies are the largest fragments of a great planet, which once revolved in an orbit about midway between Mars and Jupiter; and which, either by some internal convulsion, or external violence, has been separated into these, and probably many other smaller parts.

11. In what time does Mars revolve in its orbit?—12. What is its velocity?—13. How far is it from the sun?—14. When is Mars a morning star?—15. When an evening star?—16. When, else is it seen?—17. How many interior planets has it?—18. What are they?—19. Which is the brightest?—20. What is said of the seasons of Mars?—21. And of its light and heat?

1. By what name are Vesta, Juno, Pallas, and Ceres called?—2. Why are they called asteroids?—3. When and by whom was each of them discovered?—4. What has been supposed as to the origin of these planets?

FIRST BOOK IN ASTRONOMY.

ASTRONOMICAL AND GEOMETRICAL DEFINITIONS.

LESSON I.

The Definitions here given are in as few words as possible, with a view to their being committed to memory. Each one may be considered an answer to a distinct question, which the teacher should propose at the time of reciting. For instance, What is astronomy? What is a circle? What is a planet? &c. If this lesson is too long to be recited at one time, it can easily be divided into two or more lessons.

Astronomy is the science which describes the heavenly bodies; the Sun, Planets, Fixed Stars, and Comets.

A *planet* is an opaque, or dark body, deriving its light from another body, around which it revolves.

A *comet* is a body which moves round the sun in a very eccentric orbit or tract, and is usually accompanied with a long train of light.

The *orbit* of any celestial body is the curve or path it describes in revolving round another body.

Two celestial bodies are said to be in *opposition*, when they are in opposite points of the heavens.

Two celestial bodies are said to be in *conjunction*, when they are in the same point of the heavens.

A planet is said to move *direct*, when it appears to move according to the signs of the ecliptic.

A planet is said to move *retrograde*, when it appears to move contrary to the signs of the ecliptic.

A planet is said to be *stationary*, when it appears to remain any particular time in a certain point of the heavens.

A *circle* is a plane figure bounded by a curved line, called the circumference, every part of which is equally distant from the centre.

Outlines of Astronomy

With the improvements in telescopes, astronomy in the 19th century was an era of great discovery, including Uranus (1781), Neptune (1846), the first really accurate measurements of the vast distances to stars (1833) and work on observing the spiral nebula by astronomers including John Herschel, the author of the book we are collecting here. In addition, the techniques of physics needed to understand the observations were developing. In *Outlines of Astronomy*, John, the son of William Herschel (see page 142) summarized the current state of astronomy as a whole, as well as reported on some of his own work. It was a popular and comprehensive book, remaining in print across 11 editions until 1873 – and presumably educating many future astronomers.

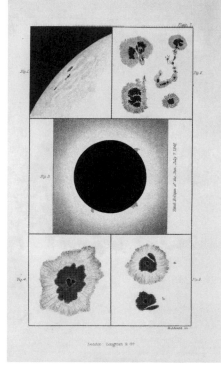

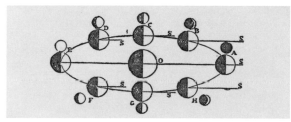

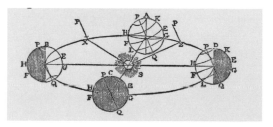

CLOCKWISE FROM TOP LEFT
Outlines of Astronomy

Drawings of two 'nebula' and a comet. From top to bottom are M13 (a globular cluster), the Comet of 1819 and the Andromeda galaxy then called the Nebula in Andromeda.

A plate with drawings of the Sun and sunspots.

Diagrams used to help explain the daily and annual motion of the Sun through the sky (right) or the phases of the Moon (left).

Telescope Teachings

A book aimed at educating younger readers, it focused on the use of telescopes to investigate the night skies. *Telescope Teachings*, published in 1859, features several beautiful illustrations by the author and included tips for objects accessible with small telescope, a review of interesting discoveries with large telescopes and implications for the understanding of astronomy. It also included details of observations of some of the recent 'Great Comets' (see pages 202 and 204–205). It was dedicated to the Earl of Rosse, who the author credits with introducing her to astronomy when he 'commended the two-inch Telescope in my possession' (i.e., he gave her a telescope).

The author, Mary Ward (1827–1869), wrote several other books related to the use of both microscopes and telescopes, which she illustrated herself. She had been born into a well-off Irish family known for their scientific interests, so despite formal education being inaccessible to girls at the time, she was scientifically educated. Her family included her cousin William Parsons (3rd Earl of Rosse), the astronomer who built 'The Leviathan of Parsonstown' with which he made the first recorded image of spiral structure in a galaxy (See pages 148 and 246). Ward showed an interest in astronomy and visited her cousin frequently – her drawings of the building process of this telescope provided a useful historical record and helped with its restoration. Alongside her books, Mary Ward has the dubious accolade of being the first person known to have been killed by a motor vehicle – in 1869, aged just 42, she was hit by an experimental steam car.

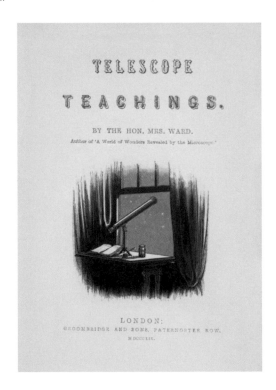

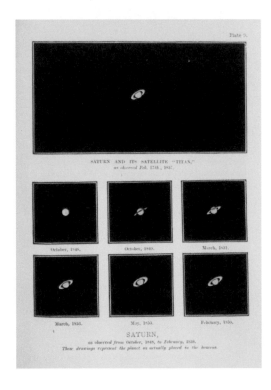

Plate 12.

HERCULES. $\beta\gamma\epsilon\lambda\alpha$ SERPENS. δ $\gamma\beta\chi\rho\varsigma$ CORONA BOREALIS.
$\epsilon\delta\gamma$ *Alphecca.*

THE COMET of 1858, (DONATI'S,)
and neighbouring stars, October 11th., 1858, at 7.15 p.m.

Telescope Teachings

Various plates to do with
observations of comets.

Details of observations of
Mercury, Venus and Mars.

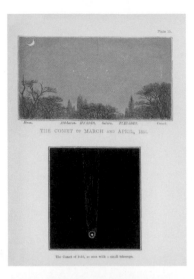

THE COMET OF MARCH AND APRIL, 1856.

The Comet of 1843, as seen with a small telescope.

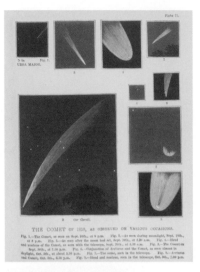

THE COMET OF 1858, AS OBSERVED ON VARIOUS OCCASIONS.

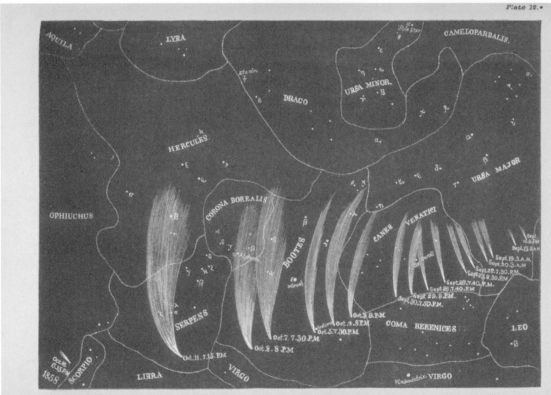

Plate 12.

Miniature Chart of the apparent Path of Donati's Comet, during the greater part of the time it was visible in the British
Isles. Intended to shew the successive changes of its position, form, and apparent dimensions during that period.

Plate 7.

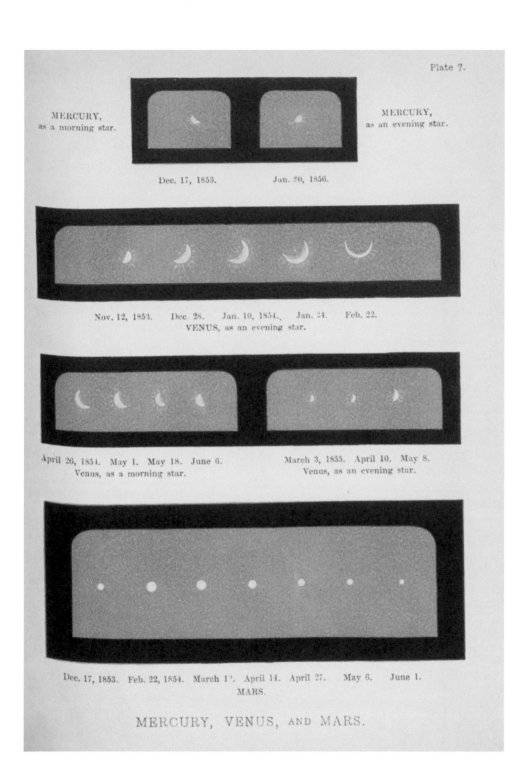

MERCURY,
as a morning star.

MERCURY,
as an evening star.

Dec. 17, 1853.   Jan. 20, 1856.

Nov. 12, 1853.   Dec. 28.   Jan. 10, 1854.   Jan. 24.   Feb. 22.
VENUS, as an evening star.

April 26, 1854.   May 1.   May 18.   June 6.
Venus, as a morning star.

March 3, 1855.   April 10.   May 8.
Venus, as an evening star.

Dec. 17, 1853.   Feb. 22, 1854.   March 12.   April 14.   April 27.   May 6.   June 1.
MARS.

MERCURY, VENUS, AND MARS.

The Story of the Heavens

An illustrated book intended for a general audience, *The Story of the Heavens* covered the basics of astronomy as understood in the late 1800s. It was written by Robert S. Ball (1840–1913) and first published in 1885. It is sometimes noted as one of the books listed in a chapter of James Joyce's *Ulysses*. Containing 27 chapters, 18 colour plates and 90 engravings, the book is both comprehensive and beautiful. It starts with a chapter on the 'The Astronomical Observatory', with some charming illustrations of contemporary telescopes. Placing it in time, alongside more typical information about the Solar System, the reader can enjoy a chapter on 'the planet of romance', not as you might expect, Venus, but a planet interior to Mercury, dubbed 'Vulcan', which at the time was much debated as a cause of oddities in the orbit of Mercury.[3] Moving outside the Solar System, the book covers 'The Starry Heavens', 'Distant Suns' and 'Star Clusters and Nebulae'. This latter chapter includes records of early observations of what we now call galaxies – these objects had not yet been determined to be outside our own Galaxy, although the chapter hints at this question. An early photograph of the Andromeda galaxy is included (by Isaac Roberts, one of the pioneers of astrophotography), and there are comments on early spectral analysis of its light suggesting it is made of stars not gas. This follows from a chapter on 'The Revelations of the Spectroscope', particularly notable in a book intended for a popular audience, as the use of astronomical spectroscopy to identify the material nature of objects in space was still very new at this point – this was cutting-edge stuff.

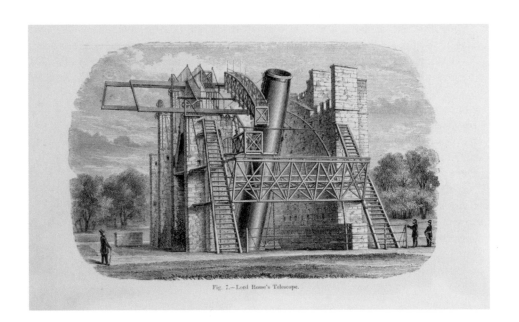

Fig. 7.—Lord Rosse's Telescope.

---

[3] Instead of 'Vulcan' we now know that the odd motion in Mercury's orbit is caused by deviations from Newtonian gravity close to the Sun – here the gravity is strong enough that Einstein's theory of general relativity is different enough from Newtonian gravity that corrections are needed to fit the orbit properly.

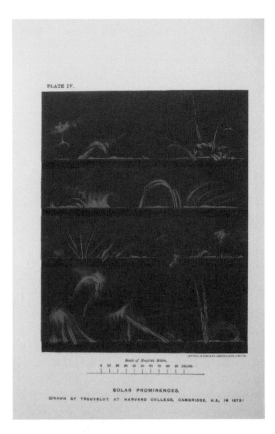

PLATE IV.

Scale of English Miles.
0 10 20 30 40 50 60 70 80 90 100,000

SOLAR PROMINENCES.

(DRAWN BY TROUVELOT AT HARVARD COLLEGE, CAMBRIDGE, U.S., IN 1873.)

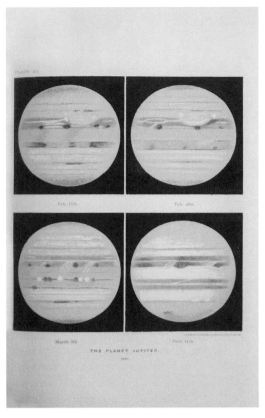

PLATE XI.

Feb. 17th.   Feb. 24th.

March 5th.   June 11th.

THE PLANET JUPITER.

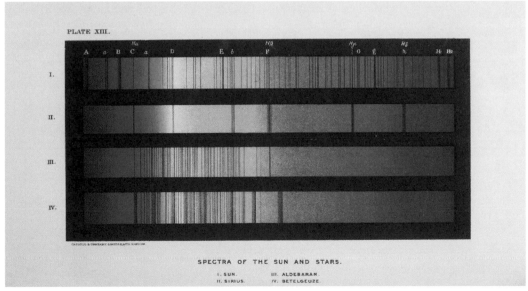

PLATE XIII.

A a B C a   D   E b   F   G g   h   H₁ H₂

I.

II.

III.

IV.

SPECTRA OF THE SUN AND STARS.

I. SUN.   III. ALDEBARAN.
II. SIRIUS.   IV. BETELGEUZE.

The Children's Book of the Stars

Here were colle&t another charmingly illustrated book that was first published in 1903 and aimed at introducing children to astronomy. The author, Geraldine E. Mitton (1868–1955), was not a professional astronomer and wrote books on a diverse set of topics – often credited as simply G.E. Mitton – mostly non-fi&tion (biographies and travel guides alongside astronomy), as well as some novels.

The Children's Book of the Stars covered the usual topics in astronomy – with much on the planets in the Solar System and some details of constellations, but it also explained the origin of tides, 'English summer and winter', 'flames from the Sun', 'the comet in the Bayeux tapestry' and had chapters on topics to do with the physics of light – how it bends when it enters water – and what we learn from astronomical spe&troscopy. Further indicating its moment in astronomical history, along with drawings, this book contains at least two astronomical photographs (the Moon and the 'Great Nebula in Andromeda') and the map of Mars shows Schiaparelli's 'canals' (see page 96) with a discussion of whether they are real or not, and if they indicate life on Mars.

poison the air. The sea tides scour our coasts day by day with never-ceasing energy, and they send a great breath of freshness up our large rivers to delight many people far inland. The moon does most of this work, though she is a little helped by the sun. The reason of this is that the moon is so near to the earth that, though her pull is a

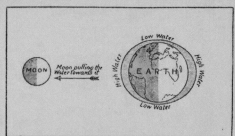

THE MOON RAISING THE TIDES.

comparatively small one, it is very strongly felt. She cannot displace the actual surface to any great extent, as it is so solid; but when it comes to the water she can and does displace that, so that the water rises up in answer to her pull, and as the earth turns round the raised-up water lags behind, reaching backward toward the moon, and is drawn up

appears like a half or three-quarter moon, as we only see a part of the side from which the sunlight is reflected. She shines like a little silver lamp,

DIFFERENT PHASES OF VENUS.

excelling every other planet, even Jupiter, the largest of all. If we look at her even with the naked eye, we can see that she is elongated or drawn out, but her brilliance prevents us from seeing her shape exactly; to do this we must use a telescope.

4—2

THE ENGLISH SUMMER (LEFT) AND WINTER (RIGHT).

The Children's Book...

A diagram explaining the cause of tides.

A clever illustration showing the phases of Venus.

(Left): This eye-catching diagram doesn't seem to explain much about the seasons at first glance – but notice the pole caps indicating the tilt of Earth and showing northern, summer (left) and winter (right).

(Below): Two more interesting and informative pages from the book showing the spectra of the Sun compared to the bright blue (and we know today much hotter) star Sirius, and a selection of constellations near Orion.

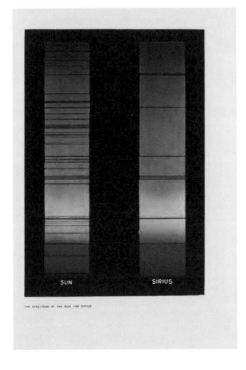

SUN   SIRIUS

THE SPECTRUM OF THE SUN AND SIRIUS.

THE TWINS

THE BULL

Aldebaran

Procyon

Betelgeuse

LITTLE DOG

ORION

Rigel

Sirius

THE DOG

THE HARE

ORION AND HIS NEIGHBOURS.

The Heavens and Their Story

The Heavens and Their Story was a book intended to both inspire and capture public imagination about astronomy. The author, Annie Maunder (1868–1947), was a British astronomer best known for her work on the Sun. She was an early expert in solar photography, observing sunspots and going on eclipse expeditions with her husband Walter Maunder (1851–1928), who was also an astronomer. Their observations and data visualizations of the locations of sunspots (such as in the 'butterfly' diagram – see opposite), and other data on solar activity (at the time mostly published under Walter's name) were a significant contribution to understanding this 11-year solar cycle, and the 'Maunder Minimum' (a prolonged absence of sunspots in the 17th and 18th centuries) is named for them.

The Heavens and Their Story was published in 1908, and although the book is credited to both Annie and Walter Maunder, the preface by Walter explains that it was really almost entirely written by Annie; the book also includes several of her own astronomical photographs, especially of the Sun, as well as the Andromeda nebula. Divided into 'stories' told by the various phenomena in the sky, more than half of it focuses on 'Stories Told by the Sun' – no surprise, given the authors' expertise as solar astronomers.

The title page boasts that the book contains 'eight coloured plates and thirty astronomical photographs and fifty-one other illustrations'. Collecting this in our Astronomers' Library helps capture a moment when photography was beginning to overtake illustration and drawing as the method to display the beauty of the heavens.

RIGHT

The Heavens and Their Story

An image of the Sun, and a details of a sunspot. Both images presumably by Annie Maunder (all other photographs are credited to the astronomer who took them, so my assumption is any without credit were due to Maunder).

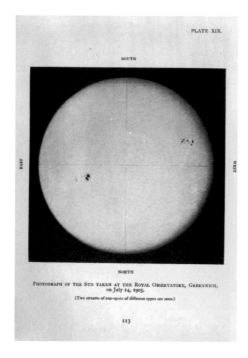

PLATE XIX.

SOUTH

EAST

WEST

NORTH

PHOTOGRAPH OF THE SUN TAKEN AT THE ROYAL OBSERVATORY, GREENWICH, on July 14, 1905.

(Two streams of sun-spots of different types are seen.)

113

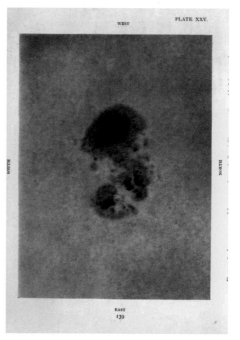

WEST

PLATE XXV.

SOUTH

NORTH

EAST
139

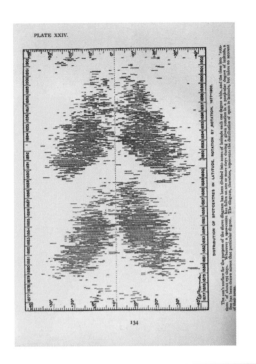

PLATE XXIV.

DISTRIBUTION OF SPOTCENTRES IN LATITUDE. ROTATION BY ROTATION, 1877-1902.

134

151

Telescopic View of the S.W. Quadrant of the Corona, May 28, 1900.
(*Drawn by Miss Lilian Martin-Leake.*)

PLATE XXVIII.

ABOVE LEFT, ABOVE RIGHT, LEFT

The Heavens and Their Story

The famous 'butterfly' diagram by Annie and Walter Maunder. It shows how the latitude of sunspots varies with time during the 11-year solar cycle – sunspots initially appear at high latitudes, then move closer and closer to the equator until the minimum occurs and the pattern resets.

A photograph of Annie Maunder with two cameras (taken by 'Miss Edith Maunder', presumably one of Walter's daughters from his previous marriage), from the magazine *Knowledge*, 1900.

A drawing of a telescopic view of the Sun showing some dramatic solar prominences.

MODERN
ASTRONOMY

O ver the last 100 years, astronomy has changed a lot; both how we do astronomy and our understanding of the universe have shifted enormously. Most historical astronomers might not even recognize the field.

By the 1920s it was well understood that the universe was bigger than the Solar System, that the 'fixed stars' were actually objects like our Sun – giant spheres of hot hydrogen and helium burning via nuclear fusion and scattered through space, collected in a huge structure we call the Galaxy (or Milky Way, after its appearance as a faint band across our skies). But debate was ongoing as to the nature of the universe beyond that. The measurement of distances to the so-called 'spiral nebulae' (done by Edwin Hubble in the 1920s, using a method developed by Henrietta Leavitt) demonstrated just how vast the universe was, and the observation that almost all of these galaxies are moving away from us (mostly due the astronomer Vesto Slipher, although interpreted by Edwin Hubble) revealed an expanding universe which started in a hot, dense 'Big Bang'.

Innovations in photography in the middle of the 19th century had started a revolution in astronomy. Rather than working long hours at night looking through a telescope by eye, astronomers collected images (or spectra) of objects in the night skies that could be analysed later. This shift opened up the field to people who had not previously had access to it, and in common with other sciences, astronomy also became increasingly professionalized.

Starting in the 1980s, astronomers were early adopters of digital cameras (or 'charge-coupled-devices'), enabling much more astronomical data to be processed much more efficiently with computers. Today's professional astronomers rarely travel to a telescope, and instead observational astronomy has turned into a field of (very big!) data science as

astronomers all over the world are able to access astronomical data over the internet. We have put massive optical telescopes, many feet across,[1] on the top of the tallest mountains on Earth and even launched them into space, and astronomers have also moved to types of light outside the visible range, building massive radio telescopes or launching sensitive high energy UV, X-ray and gamma-ray detectors. Astronomers have even reached outside of the spectrum of light: in 2015 the first gravitational wave signal was detected, revealing the in-spiral and merger of two massive black holes.

Meanwhile we can't forget the space race. Humanity started to travel to the heavens with the launch of the first satellite in 1957, the first person in space in 1961 and the first person to walk on the Moon in 1969. Robotic missions have explored all of the known planets, and the Voyager 1 and 2 missions left the Solar System (for several definitions of the edge). Today, even space is getting crowded – at least space near Earth. Since 2 November 2000, there has always been a human in space – the International Space Station has been continuously inhabited for over 20 years, and at the time of writing there are over 7,000 artificial satellites in Earth orbit, with more and more launched every year.

The communication of new astronomy, as in most scientific fields, has also changed enormously. Rather than publishing their latest research in books, most scientists write short letters or articles to journals using

LEFT AND OPPOSITE

Space and time

The cover of Stephen Hawking's classic global best-seller *A Brief History of Time*.

The NASA/ESA Hubble Space Telescope image of the star cluster Westerlund 2 and its surroundings.

highly specialized and technical language. Only a few astronomers write books. They continue to be popular to share the knowledge more widely, either through textbooks, or so-called popular-science books aimed at informal education or entertainment. And, of course, digital images of the night skies which are now easily accessible for illustration are stunning. In this section, we'll collect just a small selection of the many books which have been published about astronomy in the last 100 years.[2] We make no attempt to claim this as a definitive list. It is unapologetically a mixture of best-sellers, scientifically significant books and author favourites.

[1] At the time of writing, the largest operational telescopes have mirrors roughly 33 feet (10 m) across; 98-foot (30 m) diameter telescopes are being planned.

[2] Books are ordered roughly chronologically by publication. Where they existed in multiple editions, or if the author published multiple books in the list, we collect them at the first date of publication.

Amateur Telescope Making

Written by Albert G. Ingalls (1888–1958), the first volume was published in 1926, the last in 1953. This series of three books by the then-*Scientific American* editor and amateur astronomer is credited with popularizing telescope making in the USA.

The Internal Constitution of the Stars and *The Expanding Universe: Astronomy's 'Great Debate'*

Arthur Eddington (1882–1944) was an English astronomer noted for many contributions to both theoretical and observational astrophysics. He did early work on the internal structure and properties of stars, led expeditions that confirmed gravitational lensing around the Sun (during the 1919 eclipse) and confirmed Einstein's model of gravity and more.

In *The Internal Constitution of the Stars* (1926), the Eddington laid the foundation for modern understanding of what is going on inside stars.

The second book of his we collect, *The Expanding Universe: Astronomy's 'Great Debate'*, was published in 1933. It was an early attempt to explain to a popular audience the then very recent observations which prove 'spiral nebula' are galaxies external to our own, and that the universe is expanding. The text of the book was based on a public lecture given by Eddington at a meeting in 1932.

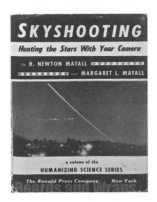

Skyshooting: Hunting the Stars with your Camera

A book of tips for amateur astronomers wanting to take images of the night skies using photography, written by R. Newton Mayall and Margaret L. Mayall in 1949.

Interplanetary Flight: An Introduction to Astronautics

A short introduction to the mathematics and physics of interplanetary flight by the noted sci-fi author Arthur C. Clarke. First published in 1950, it was intended to be accessible to general readers, including notes on likely destinations in the Solar System in addition to details of rocket science and orbital mechanics.

Introduction to Astronomy

Cecilia Payne-Gaposchkin (1900–1979) is said to have written the most brilliant PhD thesis in astronomy ever. In that work she demonstrated that all stars (and hence most of the universe) are made of mostly hydrogen and helium. Throughout her career she published several books. Here we collect her fascinating *Introduction to Astronomy* from 1954.

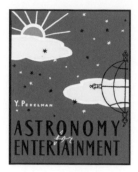

Astronomy for Entertainment

A 1954 book by the noted Russian science writer Yakov Perelman (1882–1942) with a delightfully retro-styled cover and many charming illustrations.

The Hubble Atlas of Galaxies

A collection of photographic images of galaxies by Edwin Hubble, published in 1962, after Hubble's death, by the astronomer Allan Sandage (1926–2010).

Gravitation

Written by Charles W. Misner, Kip S. Thorne and John Archibald Wheeler and first published in 1973, this is the comprehensive text that explains Albert Einstein's theory of general relativity for advanced students of the subject. It is so complete and large that it has inspired generations of students with jokes about how it has its own gravity.

The First Three Minutes: A Modern View of the Origin of the Universe

The First Three Minutes came out in 1977 and describes the physics of the very early universe to a general audience. Written by theoretical physicist Steven Weinberg (1932–2021), it remains popular for hitting a tone that is neither oversimplifying nor condescending or too technical.

Cosmos and Pale Blue Dot: A Vision of the Human Future in Space

Carl Sagan (1934–1996) was an American astronomer and highly successful popularizer of science who was particularly well known for his 1980s television series *Cosmos*, which was accompanied by a 1980 book of the same name. In all he wrote more than ten books, including fiction (*Contact*, which was turned into a film). *Pale Blue Dot* (1994), advertised as the sequel to *Cosmos*, is one of the best-selling astronomy books ever published.

A Brief History of Time: From the Big Bang to Black Holes

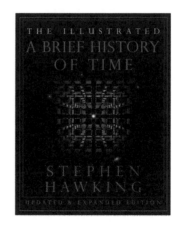

A 1988 book on cosmology and the fundamental physics of the universe, by the notable theoretical physicist Stephen Hawking (1942–2018). Hawking is perhaps as famous for his work on theoretical physics as he is for his science writing, all done while living with a debilitating and chronic disease. This book sat on the best-seller lists for a long time in the 1990s, and it inspired many jokes that most of the people who bought it never read it. I definitely did – as a teenager developing an interest in astronomy – and was gifted a copy of the 1996 illustrated version shown here as a prize in physics.

Galactic Dynamics

While they look pristine and unchanging, galaxies are highly dynamical systems – just on timescales undetectable to humans. This textbook is the definitive explanation of the classical mechanics needed to explain these complex and fascinating systems. Written by James Binney and Scott Tremaine and published in 1988.

Black Holes and Time Warps: Einstein's Outrageous Legacy

Written by astronomer Kip S. Thorne – known for his work as a consultant on the film *Interstellar* – and published in 1994, this book provides a history of developments in the theory of general relativity intended for a general audience, with a tour of some of the more outrageous consequences – like worm holes and time dilation.

A Man on the Moon: The Voyages of the Apollo Astronauts

The definitive historical account of the Apollo missions. Written by Andrew Chaikin and published in 1994.

An Introduction to Modern Astrophysics

A comprehensive textbook on modern astrophysics for the advanced student that was first published in 1995. This is the book I was tested on in 'qualifying exams' for my PhD in the early 2000s. By Bradly W. Carroll and Dale A. Ostlie.

Longitude and The Glass Universe: How the Ladies of the Harvard Observatory Took the Measure of the Stars

Dava Sobel is an American science writer and author of numerous non-fiction books related to astronomy. We'll collect two of here, 1995's *Longitude*, a story of solving the longitude problem – how to figure the longitude of a ship at sea – which was one of the big mysteries (and fundraisers) of 18th-century astronomy (see page 214). And also *The Glass Universe* from 2016, which is the story of the female astronomers at the Harvard Observatory in the early 19th century and details their contributions to the understanding of the universe.

Thread of the Silkworm

Iris Chang's 1996 story of Tsien Hsue-shen (also Qian Xuesen; 1911–2009), a Chinese–American man and pioneer of the American space programme, who was detained by the USA in 1951 under suspicion of being a communist. After returning to China in 1955, he went on to found the Chinese space programme.

Foundations of Astrophysics

A popular undergraduate textbook in astrophysics, which provides a comprehensive introduction to the topic. By Barbara Ryden and Bradley M. Peterson, published in 2009.

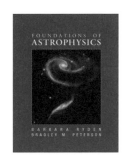

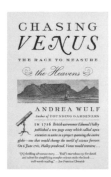

Chasing Venus: the Race to Measure the Heavens

Andrea Wulf's 2012 historical adventure story of the astronomers who travelled to time the 1761 and 1769 transits of Venus and measure the scale of the universe (see pages 99 and 210).

Bang!: The Complete History of the Universe

A rock star, the longest-ever serving television presenter and an Oxford professor of astrophysics – not the start of a joke, but the authorship of this accessible, illustrated history of the entire universe from 2016. Written by Brian May, Patrick Moore and Chris Lintott.

Hidden Figures: The American Dream and the Untold Story of the Black Women Mathematicians Who Helped Win the Space Race

The story of Black female mathematicians working at NASA who calculated the orbits for the Apollo Moon missions in the 1960s and 70s. By Margot Lee Shetterly and published in 2016 – a movie based on this book came out the same year.

Astronomy

It is said that a quarter of a million college students take a beginner's astronomy course in the USA each year, and many of them are expected to have a textbook. In the 21st century, even textbooks are going online – Openstax publish popular open source (free) online texts; this one aimed at Astro101. Featuring multiple contributing authors, it came out in 2016 but is frequently updated.

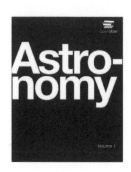

StarTalk: Everything You Ever Need to Know About Space Travel, Sci-Fi, the Human Race, the Universe, and Beyond

A book based on the popular podcast and *National Geographic* show featuring the exuberant and popular astronomy – and science in general – popularizer Neil deGrasse Tyson. Written with Charles Liu and Jeffrey Simmons, published in 2016.

A Galaxy of Her Own

By the British physicist and Head of Space Explorations for the UK Space Agency, Libby Jackson. This highly illustrated book aimed at young girls tells a series of amazing stories about women who have been involved in space travel, from the early days to those currently working in the field. Published in 2017.

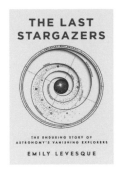

The Last Stargazers

Astronomer Emily Levesque's story of telescopes and observatories. This book which includes anecdotes of observing trips is themed around the fact that most professional astronomers rarely visit observatories anymore. It came out in 2020.

The End of Everything (Astrophysically Speaking)

A book, published in 2020, about current ideas on what will happen at the end of the universe, by the cosmologist Katie Mack.

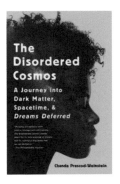

The Disordered Cosmos

A 2022 book that is partly about cosmology (especially dark matter) and particle physics and partly a biography of Chanda Prescod-Weinstein, one of the first Black women to earn a PhD in theoretical astrophysics.

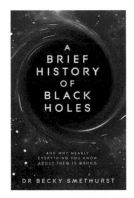

A Brief History of Black Holes

A popular science introduction to many people's favourite – and perhaps the most misunderstood astrophysical object – the black hole. By astronomer and YouTuber Becky Smethurst, 2022.

BIBLIOGRAPHY

1. Star Atlases

Cheongsang Yeolcha Bunyajido
Al Sufi, Abd Al-Rahman, *The Book of the Images of the Fixed Stars*
Hyginus, Casius Julius, Poetic Astronomy
Piccolomini, Alessandro, *De la sfera del monde, De le stelle fisse*
Bayer, Johann, *Uranometria*
Cellarius, Andreas, *Harmonia macrocosmica*
Hevelius, Elisabeth and Hevelius, Johannes, *Firmamentum sobiescianum sive uranographia*, also: *Machinae coelestis: pars prior*
Flamsteed, John, *Atlas coelestis,* also *Catalogus britannicus*

2. Mapping Other Worlds

Hevelius, Johannes, *Selenographia*
Manilius, Marcus, *The Astronomica*
Sherburne, Sir Edward, *The Sphere of Marcus Manilius*
Proctor, Richard, *The Moon: Her Motions, Aspect, Scenery and Physical Condition, Other Worlds Than Ours*
Nasmyth, James, and Carpenter, James, *The Moon: Considered as a Planet, a World and a Satellite*
Tsukioka, Yoshitoshi, *One Hundred Aspects of the Moon*
Velarianus, Petrus, *Compendium in sphaeram*
Balde, Jakob, *De eclipsi solari*
von Oppolzer, Theodor Ritter, *Canon der finsternisse*
Espenak, Fred, *Road Atlas for the Total Solar Eclipse of 2024*
Huygens, Christiaan, *Systema saturnium,* also *Cosmothereous*
Schiaparelli, Giovanni, *La vita sul pianeta Marte*
Lowell, Percival, *Mars and Its Canals*, also *Mars as the Abode of Life, Mars*
Chary, Chintamanni Ragoonatha, *Transit of Venus*
Arthusius, Gothard, *Cometa orientalis*
Gadbury, John, *De cometis*
Halley, Edmond, *A Synopsis on the Astronomy of Comets*
Weiss, Edmund, *Atlas der sternenwelt*

3. Astronomy and Culture

al-Din Abu al-Abbas Ahmad ibn al-Hajj al-Amin al Tawathi al-Ghalawi, Nasir, *The Important Stars Among the Multitude of the Heavens*
Ibn Muhammad al-Qazwini, Zakariya, *The Wonders of Creation*
Al-Qabisi, *Treatise on the Principles of Judicial Astronomy Vedanga jyotisha*
Aryabhata, *Aryabhatiya*
Somayaji, Nilakantha, *Tantrasamgraha*
Varahamihira, *Brihajjataka*
The Madrid Codex
The Dresden Codex
de Sahagún, Friar Bernardino, *General History of the Things in New Spain* (*The Florentine Codex*)
Isidore of Seville, *De responsione mundi et de astrorum ordinatione*
Bede the Venerable, *De temporum ratione*
The Book of Silk (The Divination by Astrological and Meteorological Phenomena)
Dunhuang Star Chart
Jokyo Calendar
Nobutake, Baba, *Introduction to the Study of Astronomy*
Joken, Nishikawa, *Tenmon Giron*

4. Developing Our Model of the Universe

Aristotle, *De caelo et mundo*
Scheiner, Christoph, *Disquisitiones mathematicae*
Aristarchus of Samos, *On the Sizes and the Distances of the Sun and Moon*
Ptolemy, Claudius, *The Almagest*
Apianius, Petrus, *Astronomicum caesarium,* also *Cosmographia*
Copernicus, Nicolaus, *De revolutionibus orbium coelestium*
Kepler, Johannes, *Astronomia nova, The Astronomica, Harmonice mundi, Mysterium cosmographicum*
Brahe, Tycho, *De mundi aetherei*
Blaue, Joan, *Atlas Maior*

Galilei, Galileo, *Dialogo sopra i due massimi sistemi del mondo, Discorso delle comete, Istoria e dimostrazioni intorno alle macchie solari, Sidereus nuncius*

Fludd, Robert, *Utriusque cosmi historia*

Kircher, Athanasius, *Ars magna lucis et umbrae*

Cellarius, Andreas, *Harmonia macrocosmica*

Doppelmayr, Johann Gabriel, *Atlas novus coelestis in quo mundus spectabilis*

Newton, Isaac, *Principia naturalis principia mathematica*

du Châtelet, Émilie, *Principes mathématiques*

Lepaute, Jean-André and Lepaute, Nicole-Reine, *Traité d'horlogerie*

Transit of Venus (pamphlet)

Bails, Benito, *Elementos de matematica*

Lalande, Joseph Jérôme Lefrançois

de Lalande, Marie-Jeanne
Abrégé de navigation historique théorique et pratique avec tables horaires

Somerville, Mary, *The Connection of the Physical Sciences, Mechanism of the Heavens*

5. Astronomy for Everyone

Sacrobosco, *De sphaera mundi*

Le livre de la proprieté des choses

Ferguson, James, *Astronomy Explained upon Sir Isaac Newton's Principles*

Bryn, Margaret, *A Compendius System of Astronomy*

Aspin, Jehoshaphat, *Urania's Mirror*

Blake, Reverend John Lauris, *First Book in Astronomy*

Herschel, John, *Outlines of Astronomy*

Ward, Mary, *Telescope Teachings*

Ball, Robert S., *The Story of the Heavens*

Mitton, Geraldine E., *The Children's Book of the Stars*

Maunder, Annie, and Maunder, Walter, *The Heavens and Their Story*

6. Modern Astronomy

Ingalls, Albert G., *Amateur Telescope Making*

Eddington, Arthur, *The Expanding Universe: Astronomy's 'Great Debate'*

Mayall, Margaret L. and Mayall, R. Newton, *Skyshooting: Hunting the Stars with your Camera*

Clarke, Arthur C., *Interplanetary Flight: An Introduction to Astronautics*

Payne-Gaposchkin, Cecilia, *Introduction to Astronomy*

Perelman, Yakov, *Astronomy for Entertainment*

Sandage, Allan, *The Hubble Atlas of Galaxies*

Misner, Charles W., Thorne, Kip S., Wheeler, John Archibald, *Gravitation*

Weinberg, Steven, *The First Three Minutes: A Modern View of the Origin of the Universe*

Sagan, Carl, *Cosmos, Pale Blue Dot: A Vision of the Human Future in Space*

Hawking, Stephen, *A Brief History of Time*

Binney, James and Tremayne, Scott, *Galactic Dynamics*

Thorne, Kip S., *Black Holes and Time Warps: Einstein's Outrageous Legacy*

Chaikin, Andrew, *A Man on the Moon*

Carroll, Bradly W. and Ostlie, Dale A., *An Introduction to Modern Astrophysics*

Sobel, Dava, *Longitude, The Glass Universe*

Chang, Iris, *Thread of the Silkworm*

Ryden, Barbara and Peterson, Bradley M., *Foundations of Astrophysics*

Wulf, Andrea, *Chasing Venus*

Lintott, Chris, May, Brian and Moore, Patrick, *Bang!: The Complete History of the Universe*

Shetterly, Margot Lee, *Hidden Figures*

Openstax, *Astronomy*

deGrasse Tyson, Neil, Liu, Charles and Simmons, Jeffrey, *StarTalk*

Jackson, Libby, *A Galaxy of her Own*

Levesque, Emily, *The Last Stargazers*

Mack, Katie, *The End of Everything (Astrophysically Speaking)*

Prescod-Weinstein, Chanda, *The Disordered Cosmos*

Smethurst, Becky, *A Brief History of Black Holes*

INDEX

A

Ach Valley tusk fragment 18
Adams, John Couch 218
Akiyama Buemon 76
al-Din Abu al-Abbas Ahmad ibn al-Hajj
    al-Amin al-Tawathi al-Ghalawi,
    Nasir 112
  *The Important Stars Among the
    Multitude of the Heavens* 112,
    113, 225
Alexandria 152, 156
Alpha Orionus 30
Al-Qabisi 116
  *Treatise on the Principles of Judicial
    Astronomy* 116
Al Sufi, Abd Al-Rahman 22
  *The Book of the Images of the Fixed
    Stars* 22
Andromeda Galaxy 92, 241, 246, 248,
    250. *See also* constellations
Apianius, Petrus 158
  Astronomicum caesarium 158, 159, 162
Apollo 17 107
Aries (constellation) 22
Aristarchus of Samos 151
  *On the Sizes and the Distances of the
    Sun and Moon* 151
Aristotle 142, 150, 155, 177
  De Caelo et Mundo 150
Arthusius, Gothard 101
  Cometa Orientalis 101, 102
Aryabhata 119
  Aryabhatiya 56, 119
Aspin, Jehoshaphat 238
  Urania's Mirror 227, 238, 239
asterism 50
  Big Dipper/Plough 18, 50
asteroid 10, 72, 240
  Juno 240
  Pallas 240
  Vesto 240
astrolabe 21, 158
Astronomer Royal 45, 49, 103, 202
astronomy 8, 10, 12, 14, 18, 28, 30, 41,
    42, 49, 58, 62, 66, 67, 84, 88,
    98, 103, 106, 107, 109, 110,
    111, 114, 116, 118, 119, 122,
    125, 126, 127, 128, 130, 135, 136,
    137, 138, 142, 150, 152, 155, 156,
    158, 164, 168, 171, 172, 180, 186,
    192, 200, 202, 208, 212, 214, 215,
    222, 224, 225, 228, 233, 235, 236,
    240, 241, 242, 246, 248, 250, 254,
    255, 257, 259, 260, 262
Astronomy and Culture 104–138
Astronomy for Everyone 220–250
astrophysics 103, 201, 256, 260, 261, 263
Auriga 41, 43
autumn 107, 202
autumnal equinox 107, 109
Azophi. *See* Al Sufi, Abd Al-Rahman

B

Bails, Benito 212
  Elementos de Matematica 212
Balde, Jakob 82
  De eclipsi solari 82
Ball, Robert S. 246
  The Story of the Heavens 246
Bayer, Johann 30
  Uranometria 30, 31, 33, 34, 35, 41
Bayeux Tapestry 100
Bede the Venerable 134
  De Temporum Ratione 134
  Historia Ecclesiastica 134
Beer, Wilhelm 71
Belize 122
Betelgeuse. *See* Alpha Orionus
Bible, the 38, 134
Big Bang 148, 254, 259
Big Dipper 18, 50
Binney, James 259
  Galactic Dynamics 259
black-drop effect 211
black holes 8
Blake, Reverend John Lauris 240
  First Book in Astronomy 240
Blaue, Joan 174
  Atlas maior 174
Book of Silk, The (*The Divination by
    Astrological and Meteorological
    Phenomena*) 136
The Book of the Properties of Things 230
Brahe, Tycho 12, 34, 48, 103, 168, 169, 170,
    172, 173, 187, 189
  De mundi aetherei 172
Browning, John 10, 92
Bryn, Margaret 236
  A Compendius System of Astronomy 236

C

Cai Lun 136
calculus 200, 202, 206, 216, 217, 218
calendar 106, 109, 110, 114, 118, 122,
    125, 128, 135, 137
  Hindu calendar 118
  Islamic calendar 109
  Jewish calendar 109, 135
Cambridge Observatory 14
Carpenter, James 72
  *The Moon: Considered as a Planet, a
    World and a Satellite* 72
Carroll, Bradly W. 260
  Introduction to Modern Astrophysics 260
Cassini, Giovanni 93
Cassiopeia (constellation) 23, 26, 48
Cellarius, Andreas 35, 166, 188
  Harmonia macrocosmica 35, 188, 189,
    191
Ceres 240
Chaikin, Andrew 260
  A Man on the Moon 260
Chang, Iris 261
  Thread of the Silkworm 261
Charles V 122, 158
Chary, Chintamanni Ragoonatha 99, 225
  Transit of Venus 99
Cheomseongdae 136, 138
Cheomseongdae Observatory 138
Cheonsang Yeolcha Bunyajido 20, 21,
    136
Chichen Itza 120, 122
Chinese zodiac 137
Christmas 109
circumradius 168
Clarke, Arthur C. 257
  *Interplanetary Flight: An Introduction to
    Astronautics* 257
comets 100, 101, 103, 176, 242
  Great Comet of 1577 173
  Great Comet of 1618 101
  Great Comet of 1680 202

PICTURE CREDITS

The publishers thank the following for permission to reproduce the illustrations in this book. Every effort has been made to provide correct attributions. Any inadvertent errors or omissions will be corrected in subsequent editions.

Alamy: Historic Images / Alamy Stock Photo 9; Lebrecht Music & Arts / Alamy Stock Photo 12 (TR); Artepics / Alamy Stock Photo 13; Historic Collection / Alamy Stock Photo 18; agefotostock / Alamy Stock Photo 23; Aclosund Historic / Alamy Stock Photo 24 (TL); history_docu_photo / Alamy Stock Photo 29 (TR); The History Collection / Alamy Stock Photo 32; The History Collection / Alamy Stock Photo 33; M&N / Alamy Stock Photo 41; Heritage Image Partnership Ltd / Alamy Stock Photo 47; Heritage Image Partnership Ltd / Alamy Stock Photo 48; The History Emporium / Alamy Stock Photo 49; Album / Alamy Stock Photo 50 (B); PRISMA ARCHIVO / Alamy Stock Photo 55; GRANGER - Historical Picture Archive / Alamy Stock Photo 60; Chronicle / Alamy Stock Photo 61; The Granger Collection / Alamy Stock Photo 63 (B); Antiqua Print Gallery / Alamy Stock Photo 69 (T); The Picture Art Collection / Alamy Stock Photo 89 (R); GRANGER - Historical Picture Archive / Alamy Stock Photo 96; Heritage Image Partnership Ltd / Alamy Stock Photo 102 (T); Artokoloro / Alamy Stock Photo 135; World History Archive / Alamy Stock Photo 136; Joshua Davenport / Alamy Stock Photo 139; The Picture Art Collection / Alamy Stock Photo 143; GRANGER - Historical Picture Archive / Alamy Stock Photo 152; PRISMA ARCHIVO / Alamy Stock Photo 153; GRANGER - Historical Picture Archive / Alamy Stock Photo 154; Pictures Now; History and Art Collection / Alamy Stock Photo 173 (BR); Chronicle of World History / Alamy Stock Photo 206 (R); The Picture Art Collection / Alamy Stock Photo 211 (B); Album / Alamy Stock Photo 212; Album / Alamy Stock Photo 213 (L); Album / Alamy Stock Photo 213 (R); Pictorial Press Ltd / Alamy Stock Photo 218; Photo 12 / Alamy Stock Photo 230 (T). **Archive.org**: 80, 81 (L), 81 (TR), 109, 222, 225 (T), 232, 233, 234 (T), 234 (B), 250 (L), 250 (R), 251 (TL), 251 (B). **Bridgeman Art Library**: 26, 59. **Courtesy of the British Library Board** (Royal MS 13 A XI: f.30v, f.33v, f.143v): 134 (R), 134 (C), 134 (L). **© The Trustees of the British Museum**: 10 (TR). **Reproduced by kind permission of the Syndics of Cambridge University Library**: 200 (L), 200 (R), 201 (L), 201 (R). **David Rumsey Map Collection, David Rumsey Map Center, Stanford Libraries**: 192, 193, 194 (L), 195, 196, 197, 198-199. **e-rara.ch / Historical scientific literature (ETH Library)**: 28 (BL), 28 (BR), 29 (TL), 29 (BL), 29 (BR), 31 (T), 31 (B), 34, 65 (L), 65 (R). **e-rara.ch / ETH-Bibliothek Zürich**: 35, 30, 36, 37, 40, 42 (T), 42 (B), 43 (T), 43 (B), 44 (T), 44 (B), 150 (L), 150 (R), 151 (L), 151 (R). **e-rara.ch:** 170 (L), 170 (C), 170 (R), 183 (R), 186 (L), 186 (R), 187 (BL), 187 (BC), 187 (BR), 188, 189, 190-191, 208 (L), 208 (R), 209 (T), 209 (L), 209 (R), 214, 215 (L), 215 (R), 216 (L), 216 (R), 217 (L), 217 (R). **e-rara.ch / Universitätsbibliothek Basel**: 184 (L), 184 (R), 185, 203, 204-205, 206 (L), 207 (T), 207 (B). **© NASA, ESA, A. Simon (Goddard Space Flight Center), and M. H. Wong (University of California, Berkeley) and the OPAL team**: 89 (TL). **ESO, IAU and Sky & Telescope**: 51 (B). **Courtesy of Fred Espenak**: 87, 87 (TR). **Getty Images**: Science & Society Picture Library / Getty Images: 10 (BL), 12 (TL), 15, Getty Images / J. W. Draper / Stringer: 68, Getty Images / ullstein bild Dtl. / Contributor: 131 (B). **Courtesy of HathiTrust**: 82, 83 (L). **Holden Blanco**: 8. **Lessing J. Rosenwald Collection, Rare Book and Special Collections Division of the Library of Congress, Washington, DC**: 10 (TL). **Library of Congress, World Digital Library**: 22, 24 (TR), 24 (BL), 24 (BR), 25, 27 (TL), 27 (TR),27 (BL), 27 (BR), 54, 58 (T), 58 (B), 62, 63 (T), 64. **Library of Congress**: 72 (L), 72 (R), 73 (TL),73 (B), 74 (TL), 74 (TR), 74 (BL), 74 (BR), 75 (T), 75 (B), 118 (R), 155, 156-157, 164, 165, 223, 224, 241 (TL), 241 (TR), 241 (BL), 241 (BR). **National Diet Library / World Digital Library**: 76 (L), 76 (R), 77, 78, 79, 92, 93 (L), 93 (R), 94-95, 97 (TR), 97 (BL). **Library of Congress, Mamma Haidara Commemorative Library**: 112 (L), 112 (R), 113, 116, 117 (T). **Library of Congress / Qatar National Library / World Digital Library**: 117 (B), Library of Congress / Saxon State and University Library, Dresden / World Digital Library: 123 (L), 123 (R), 124 (L), 124 (R), 125 (L), 125 (CL), 125 (CR), 125 (R), **Library of Congress / Medicea Laurenziana Library, Florence / World Digital Library**: 126, 127 (L), 127 (R), 128 (L), 128 (C), 129. **Library of Congress, Rare Book and Special Collections Division**: 131 (L), 131 (R), 132 (TL), 132 (TR), 132 (BL), 132 BR, 133, 172, 173 (TL), 173 (TR), 176 (L), 176 (R), 177 (L), 177 (R), 178 (L), 178 (R), 179, 181, 182 (L), 182 (R), 183 (L). **Courtesy of The Linda Hall Library of Science, Engineering & Technology**: 225 (B), 236 (T), 236 (B), 237 (TL), 237 (TR), 237 (BL), 237 (BR), 240 (T), 240 (BL), 240 (BC), 240 (BR), 242 (L), 242 (R), 243, 244 (TL), 244 (TR), 244 (B), 245, 246, 247 (TL), 247 (TR), 247 (B). **The Metropolitan Museum of Art**: 158, 159 (T), 159 (BL), 159 (BR), 160-161, 162, 163, 228, 229 (T), 229 (B). **Milestone Books**: milestone-books.de: 69 (B), 70 (T), 70 (B), 71 (T), 71 (B), 138 (L), 138 (C), 138 (R). **NASA, ESA, CSA, and STScI**: 4, 108, 149. **NASA**: 57, 106, 107, 194 (R). **NASA/GSFC/ Arizona State University**: 73 (TR). **NASA and The Hubble Heritage Team (STScI/AURA)**: 91 (T). **NASA/SDO, AIA**: 98. **NASA, ESA, S. Beckwith (STScI) and the Hubble Heritage Team (STScI/AURA)**: 146-147. **NASA/Moore Boeck**: 175. **NASA/JPL**: 219. **NASA, ESA, the Hubble Heritage Team (STScI/AURA), A. Nota (ESA/STScI), and the Westerlund 2 Science Team**: 255. **National Library of Australia**: 45, 46, 50 (T). **From The New York Public Library**: 169 (T), 169 (TR), 169 (BR). **PBA Galleries**: 66, 67 (T), 67 (B). **Project Gutenberg**: 248 (L), 248 (R), 249 (T), 249 (BL), 249 (BR). **Qatar Digital Library**: 114 (L), 114 (R), 115. **Shutterstock**: Granger / Shutterstock 38, 39, 166–167; Xinhua/Shutterstock 110; Nicolas Economou/NurPhoto/Shutterstock 111; Joe Pepler/Shutterstock 120-121; Universal History Archive/Shutterstock 121 (T); Shutterstock / ds_us 187 (T). **Smithsonian**: 103 (L); 103 (R). **Science Photo Library**: ETH-BIBLIOTHEK ZÜRICH / SCIENCE PHOTO LIBRARY 89 (BL), 90 (T), 90 (B); ROYAL ASTRONOMICAL SOCIETY / SCIENCE PHOTO LIBRARY: 99; DETLEV VAN RAVENSWAAY / SCIENCE PHOTO LIBRARY: 102 (B); STEVE ALLEN / SCIENCE PHOTO LIBRARY 118 (L); ROYAL ASTRONOMICAL SOCIETY / SCIENCE PHOTO LIBRARY 148, 210; RIJKSMUSEUM / SCIENCE PHOTO LIBRARY 211 (T). **Unsplash**: / Jonathan Francisca 11, / Vincentiu Solomon 6–7. **Wellcome Collection**: 101 (L), 101 (R), 229 (R), 235. **Wikimedia Commons**: 20, 56, 84, 85, 86, 91 (B), 99, 100, 144-145; / Gabriel Seah: 19, 20; / Frank Vincentz: 21 (TL); / Khalili Collections: 21 (TR); / Etphonehome: 51 (T); / Mukerjee: 119 (T); / Ms Sarah Welch 119 (B); Jbarta: 137; / Leinad-Z-commonswiki: 142; / Stanlekub: 174-175; / Marikevanroon20: 230 (B); / DcoetzeeBot: 231; / Adam Cuerden: 226-227, 238 (L), 238 (R), 239 (T), 239 (B); / Edith Maunder: 251 (TR).

Ballantine: 259 (TL2); Bantam: 259 (R); Basic Books: 258 (BR), 261 (TL); Bloomsbury: 260 (BL1); Bold Type Books: 263 (BL); Cambridge University Press: 256 (C), 260 (CR), 261 (TR); Carlton Books: 261 (BR); Carnegie Institution of Washington: 258 (TR); Century: 262 (BL); Harper & Brothers: 257 (C); Macmillan: 263 (BR); National Geographic: 262 (LC); Openstax: 262 (TR); Pelican Books: 256 (B); Princeton University Press: 258 (BL), 259 (BL); Random House: 259 (TL1); Scribner:263 (TR); Smithsonian: 257 (B); Sourcebooks: 263 (TL); The Ronald Press Company: 257 (T); Time Life: 260 (TL); University Press of the Pacific: 258 (TL); Viking: 260 (BL2); W. W. Norton & Company: 260 (TR); William Morrow: 262 (TL); Willmann-Bell: 256 (TL), 256 (TC), 256 (TR); Windmill Books: 261 (BL).